Plant Ethics

Large parts of our world are filled with plants, and human life depends on, interacts with, affects and is affected by plant life in various ways. Yet plants have not received nearly as much attention from philosophers and ethicists as they deserve. In environmental philosophy, plants are often swiftly subsumed under the categories of "all living things" and rarely considered thematically. There is a need for developing a more sophisticated theoretical understanding of plants and their practical role in human experience.

Plant Ethics: Concepts and Applications aims at opening a philosophical discussion that may begin to fill that gap. The book investigates issues in plants ontology, ethics and the role of plants and their cultivation in various fields of application. It explores and develops important concepts to shape and frame plants-related philosophical questions accurately, including new ideas of how to address moral questions when confronted with plants in concrete scenarios.

This edited volume brings together for the first time, and in an interdisciplinary spirit, contemporary approaches to plant ethics by international scholars of established reputation. It will be of great interest to students and scholars of Philosophy and Ethics.

Angela Kallhoff is a Professor of Ethics with special emphasis on Applied Ethics at the University of Vienna, Austria.

Marcello Di Paola is a post-doctoral researcher at the University of Vienna, Austria, and Research and Teaching Fellow, Department of Political Science, LUISS University, Rome, Italy.

Maria Schörgenhumer is a PhD candidate at the University of Vienna, Austria.

Routledge Environmental Humanities

Series editors: Iain McCalman and Libby Robin

The *Routledge Environmental Humanities* series is an original and inspiring venture recognizing that today's world agricultural and water crises, ocean pollution and resource depletion, global warming from greenhouse gases, urban sprawl, overpopulation, food insecurity and environmental justice are all *crises of culture*.

The reality of understanding and finding adaptive solutions to our present and future environmental challenges has shifted the epicentre of environmental studies away from an exclusively scientific and technological framework to one that depends on the human-focussed disciplines and ideas of the humanities and allied social sciences.

We thus welcome book proposals from all humanities and social sciences disciplines for an inclusive and interdisciplinary series. We favour manuscripts aimed at an international readership and written in a lively and accessible style. The readership comprises scholars and students from the humanities and social sciences and thoughtful readers concerned about the human dimensions of environmental change.

Plant Ethics

Concepts and Applications

Edited by Angela Kallhoff,
Marcello Di Paola and
Maria Schörgenhumer

Routledge
Taylor & Francis Group
LONDON AND NEW YORK

earthscan
from Routledge

First published 2018
by Routledge

2 Park Square, Milton Park, Abingdon, Oxfordshire OX14 4RN
52 Vanderbilt Avenue, New York, NY 10017

Routledge is an imprint of the Taylor & Francis Group, an informa business

First issued in paperback 2019

British Library Cataloguing-in-Publication Data
A catalogue record for this book is available from the British Library

Library of Congress Cataloging-in-Publication Data
Names: Kallhoff, Angela, editor. | Di Paola, Marcello, editor. | Schörgenhumer, Maria, editor.
Title: Plant ethics : concepts and applications / edited by Angela Kallhoff, Marcello Di Paola, and Maria Schörgenhumer.
Description: Abingdon, Oxon ; New York, NY : Routledge, 2018. | Series: Routledge environmental humanities | Includes bibliographical references and index.
Identifiers: LCCN 2018004350 (print) | LCCN 2018011747 (ebook) | ISBN 9781315114392 (eBook) | ISBN 9781138079212 (hardback)
Subjects: LCSH: Human-plant relationships. | Plants (Philosophy) | Plants, Useful. | Gardens. | Landscapes.
Classification: LCC QK46.5.H85 (ebook) | LCC QK46.5.H85 P5245 2018 (print) | DDC 581.6/3—dc23
LC record available at https://lccn.loc.gov/2018004350

ISBN: 978-1-138-07921-2 (hbk)
ISBN: 978-0-367-85599-4 (pbk)

Typeset in Goudy
by Florence Production Ltd, Stoodleigh, Devon, UK

Contents

Illustrations

Contributors

Attfield, Robin: Professor of Philosophy at Cardiff University, UK

Blok, Vincent: Associate Professor in Sustainable Entrepreneurship, Business Ethics and Responsible Innovation, Wageningen University, the Netherlands

Coeckelbergh, Mark: Professor of Philosophy of Technology, University of Vienna, Austria

Cooper, David E.: former Professor of Philosophy, Durham University, UK (retired)

D'Angelo, Paolo: Professor of Aesthetics at the University Roma Tre, Italy

Di Paola, Marcello: Post-Doc Researcher "New Directions in Plant Ethics", University of Vienna, Austria and Research and Teaching Fellow, Department of Political Science, LUISS University, Rome

Dürnberger, Christian: Research Assistant at the Messerli Forschungsinstitut in Vienna, Austria; LMU Munich, Germany

Gremmen, Bart: Professor of Ethics in Life Sciences, Wageningen University, the Netherlands

Houle, Karen: Associate Professor of Philosophy at the University of Guelph, Canada, and adjunct graduate faculty in the Schools of Fine Art and Music and the Guelph-Humber School of Creative Writing

Kallhoff, Angela: Professor of Ethics with special emphasis on Applied Ethics at the University of Vienna, Austria

Karafyllis, Nicole: Chair for Philosophy of Science and Technology at TU Braunschweig, Germany

Krebs, Angelika: Professor of Practical Philosophy at the University of Basel, Switzerland

Mazzolai, Barbara: Coordinator of the Center for Micro-BioRobotics at the Italian Institute of Technology, Pisa, Italy

Odparlik, Sabine: Leibniz-Institute of Plant Genetics and Crop Plant Research, Gatersleben, Germany

Pellegrino, Gianfranco: Assistant Professor, Department of Political Science, LUISS University, Rome, Italy

Pirscher, Frauke: Senior Research Fellow, Department of Agricultural, Environment and Food Policy, Martin Luther University Halle-Wittemberg, Germany

Pouteau, Sylvie: Institut National de la Recherche Agronomique – INRA (National institute for Agronomic Research, Paris, France)

Salvini, Pericle: Post-Doc Research Fellow at the BioRobotics Institute of Scuola Superiore Sant'Anna, Pisa, Italy

Sandler, Ronald: Associate Professor of Philosophy, Northeastern University, USA

Schörgenhumer, Maria: Prae-Doc Researcher "New Directions in Plant Ethics", University of Vienna, Austria

Thompson, Paul B.: Professor of Agricultural, Food and Community Ethics, University of Michigan, USA

Višak, Tatjana: Researcher/Lecturer in Ethics, Department of Philosophy, Mannheim University, Germany

Acknowledgments

Scientific work is not only a collaborative undertaking, but also the endeavour to break new grounds to change scientific landscapes and to introduce new and challenging theses. Plant ethics is such an endeavour. The editors wish to thank the many open-minded guests of our conferences and the many supportive colleagues at the University of Vienna. In the course of the research project "New Directions in Plant Ethics" (P27172 – plantethics.univie.ac.at), we have profited from inspiring debates and contributions of researchers from many different contexts and backgrounds. In particular, we wish to thank the authors of the chapters of this book. They do not only share the vision of an ethics that includes plant life, but have also been particularly engaged in rethinking plant life in ethics. Our thanks also go to Michaela Bartsch, who contributed to organizing lecture series and conferences on plant ethics, to Michael Bruckner and Wolfgang Damoser for their editorial work on the manuscript, and to Leila Walker and Richard Sanders from Routledge, who supported us throughout the publication process. Finally, we wish to thank the Austrian Science Fund (FWF), whose financial support was essential for realizing this project.

Introduction

Angela Kallhoff, Marcello Di Paola and Maria Schörgenhumer

Large parts of our world are filled with plants. Plants inhabit the great outdoors, our gardens, hectares of cultivated fields, our mythologies and imagination, and the vases on our dinner tables. The existence and persistence of almost all life on Earth is premised on some central functions of plants. Every living being interacts with, affects, and is affected by plant life in various ways. Among other things, plants transform light into energy and store it, breathe oxygen into the atmosphere, synthesize products that mammals and other animals need for survival, and deliver food. Yet, plant life has traditionally been marginalized in Western philosophy. The philosophical marginalization of plants may be due to implicit allegiances to dubious ontological rankings inspired by the idea of a *Scala Naturae*, as well as to plants themselves lacking many of the properties that attract the perceptual and moral attention of humans. Plants do not make noises, they speak no language of sounds, they seemingly do not move, they do not cry when damaged, and they apparently lack self-awareness and conscious decision making capacities. This often leads us to think of plants as silent, passive, static, background entities. But we should know better.

Plant research shows that plants do move, display a range of sophisticated defense mechanisms when exposed to stress and parasites, have unparalleled adaptive capacities, and communicate in various ways. Though radically different from humans and non-human animals, plants are extremely complex living beings.

Within the field of environmental ethics, plants have been discussed but have also all too often been subsumed under the category of "all living beings" and rarely considered thematically. This is much too coarse-grained a categorization: it misses, with no deep philosophical justification, the peculiarity of a form of life – the vegetal – that displays unique features whose complexity we are only just beginning to understand. A free-standing ethics of plants is at best embryonic. In part, as already suggested, this is due to difficulties in construing plants as morally considerable. Most environmental philosophers would say that strategies to include plants within the moral universe by extensionist routes do not succeed. To our best understanding, plants are not sentient and lack self-awareness and conscious agency. Biocentric philosophers have argued that plant life is morally relevant, but there are good reasons why plant ethics would want to avoid being considered a mere implication of biocentrism. First, because biocentrism argues for the intrinsic value and moral considerability of all living beings, with no dedicated interest to plants. If there is anything that is peculiarly valuable about plant life, biocentrism is in no privileged position to acknowledge it. Second, biocentrism is not a mainstream position in environmental ethics. Both points also apply to ecocentric views that accord moral status to plants as parts of morally valuable ecological wholes.

The case of plants in environmental ethics is really peculiar. Plants are necessary, useful, beautiful, complex, diverse, and tell stories about peoples, times and places – yet no publishers or readers seem to have a problem if environmental ethics textbooks include no chapter on plant ethics. This book aims at opening a philosophical discussion that may begin to fill that peculiar gap. It investigates issues in plants ontology and ethics, and explores the role of plants and their cultivation in various fields of application. The central aim of this book is to shape and frame plants-related philosophical questions accurately, and develop new ideas on how to address plant-related moral questions when confronted with concrete scenarios, including the development of new breeding technologies and robotics in agriculture.

We acknowledge at the outset that the otherness of plant life inhibits a fully convincing application of extensionist strategies to the identification of plants' values and moral significance. But there are alternatives to extensionism. The alternatives proposed in this book are versions of three broad approaches to plant ethics: a *non-moral yet inclusive approach*; a *relational approach*; and a *value-in-nature approach*. Each of these approaches provides a normative framework for plant ethics but does not lose sight of the fact that plants differ radically from other life-forms, including humans and other animals.

A *non-moral yet inclusive approach* starts with the insight that plant life is distinctively important for the life of humans as well as non-human animals. On this view, plants have value and moral importance because they are irreplaceable elements of the good life of the sentient beings that inhabit planet Earth. This approach is ecumenical with regards to the reasons that make plants valuable and morally important (to sentient beings): it welcomes aesthetic, prudential, narrative, and scientific reasons for valuing and respecting plants, and couples these with moral reasons that refer to the good of h umans and other animals. In addition, it often speaks the language of justice: protecting forests, botanical biodiversity, plants populations, and individual plants is in many cases one way of being fair to all those humans and non-human animals that can in various ways benefit from the vegetal world today and in the future.

A *relational approach* has it that the value and moral importance of plants do not descend from plants' ontological and moral status, but rather emerge from within human practices that relate to plants. Relational approaches do not typically speak the language of duties and obligations, but of care and virtue; and they are particularly apt to the exploration of the normative criteria and reasons at work in practices of plant cultivation.

A *value-in-nature approach* emphasizes not our reasons to value plants, nor our relation to them, but rather the values of plants themselves. This approach should not be confused with a search for the intrinsic value of plants (a search that has been attempted before, with less than fully persuasive results): it is rather an exploration of concrete, empirically tested values in, of, and for plant life, including integrity and flourishing. It also includes proposals to re-adjust value talk in plant ethics so that it is not premised on an individualistic ontology of entities.

These three approaches can be developed in various ways, creatively combined, and channeled in many different directions. The otherness of plants inspires philosophical attempts that are other than mainstream, and it is one of the aims of this book to show that, just off the beaten paths, a rich conceptual and normative repertoire is available to us for reflecting on plant ethics.

This book is not the very first step in the field. Two important explorations of plant ethics have come from the German-speaking world in the early 2000s. In 2001, Werner Ingensiep delivered a voluminous study on plant life that investigated the interpretation of the "plant soul" in the history of philosophy (Ingensiep 2001). A more specifically ethical treatment of plant life, Angela Kallhoff's *Principles of Plant Ethics: the Evaluation of Plant Life in Biology and Philosophy*, offered an articulated Neo-Aristotelian perspective on plant ethics, and opened the field as it is now developed in this volume (Kallhoff 2002).

As noted earlier, in the English-speaking world philosophical interest in plants has been a feature of biocentric and some ecocentric perspectives in environmental ethics. Indeed biocentrism has often referred to plants as model entities of its concern (Attfield 1981; 1991), and spurred discussions on plants in other fields of philosophy. In legal philosophy, for example Christopher Stone (1996) has built on biocentric grounds to argue for the legal standing of trees, and flamboyantly defended legal rights that would protect plant life.

More recently, other works have contributed to a reassessment of plant life from non-biocentric perspectives and various angles in philosophy. Among these are Michael Marder (2013), Matthew Hall (2011), Stefano Mancuso and Alessandra Viola (2015), and Richard Mabey (2015). Plants have also received, and are still receiving, great attention in food ethics and agricultural ethics.

Content of the book

What sets this book apart from most other works in environmental ethics is that it presents and explores plant ethics as a field of inquiry that has its own argumentative resources at both a theoretical and a more applied level. The book is divided into two main parts: the first part presents concepts and approaches to plant ethics, while the second considers some applications.

The first three chapters introduce the reader to the meanders and strictures of plant ethics by considering core issues in axiology and reflecting on the applicability to plants of fundamental ethical concepts such as welfare and interests. We begin with a discussion on the value of plants authored by *Gianfranco Pellegrino*, who provides an overview of the main axiological issues that plant ethics should take on. He then defends a specific thesis whereby individual plants have final value, with the source of this value being what nature, humans, or possibly both, have "invested" so that these particular individual plants may come to be as they are.

Tatjana Višak considers plant ethics from a utilitarian point of view. She asks two important questions: first, is there such a thing as plant welfare? And second, how does our treatment of plants affect welfare? Višak argues that plants are not welfare subjects and that our treatment of plants affects the welfare only of animals, including humans. According to utilitarianism, a plant ethics must be indirect and derivative from animal and human ethics; we have reason to protect plants in order to promote the welfare of humans and animals.

Ronald Sandler argues that plants have inherent worth: that they have interests that moral agents ought to care about for the sake of plants themselves. This claim is supported by a teleological account of interests and an etiological account of teleology: plants do have interests because they are teleologically organized as a result of natural selection. This position also implies that bacteria and artifacts have interests, since they too have a selection etiology and are teleologically organized. Opponents of the idea that plants have inherent worth often present these implications as absurd. Sandler discusses possible responses to this fundamental "absurdity" objection to plant ethics.

Chapters 4 and 5 consider two concepts that may prove central to plants ethics: flourishing and dignity. In Chapter 4, *Angela Kallhoff* presents a neo-Aristotelian approach to the evaluation of plant life. She argues that the concept of plant "flourishing" provides a yardstick in reference to which we can objectively judge whether plants are harmed or benefited, and thanks to which we can build an ethics of plants that takes its main cues from scientific botany rather than philosophical axiology. In addition, Kallhoff underlines how the notion of flourishing helps to define moral claims in different spheres of human–nature interaction, including wild nature, cultivated nature and used nature.

In Chapter 5, *Sabine Odparlik* articulates a concept that has made its way through legislation in a number of German-speaking countries and propels an attempt to give legal status and recognition to plants: the Kantian concept of dignity. This is a notion that Kant referred to human beings. arguing that human beings have dignity primarily as and because they are rational moral agents. For this reason, the extension of the notion of dignity to plants may sound quite strained. Odparlik explains how conceptual and practical worries with the notion of dignity as applied to plants can be defused or qualified. In doing so, she also illuminates some of the fundamental difficulties in plant ethics – including the need to refer to ethical concepts constructed for kingdoms very different from the kingdom of plants.

Chapters 6 and 7 give us a taste of postmodern plant ethics. Both insist that our very ways to think about plants proceed by false analogies with the animal world, and that this is a burden that plant ethics should shake off its back. In Chapter 6, *Karen Houle* argues that the concepts that we use should be described as the body of our thought: much like bodies, these concepts are inherited, undergo morphogenesis, and in turn give rise to a patterned "being-of-thought". She explains how the normative conceptual systems that we currently employ in philosophy are dominated by binary oppositions much in the same way that the human animal body is marked by "left-right" bi-laterality. Our "becoming-ethical" with and regarding plants requires different ways of being and thinking than those promoted by our entrenched "left-right", "either-or" patterns. These ways can be discerned from the very characteristics of plants' bodily nature. In particular, the development of formal symmetries in plant cells, bodies, and parts runs through a different sequence than does the development of animal bodies and parts, with notable differences in the distribution of bi-laterality in the whole. Houle suggests that plants' bodily nature inspires non-binary thinking, and that such thinking is both necessary for plant ethics and a gift that plant ethics can make to philosophy more generally.

Sylvie Pouteau provides further crucial insights on how plants should be conceptualized for plant ethics to be both ontologically accurate and philosophically distinct. She argues that in contrast to human beings and non-human animals, plants are open, non-self-centered subjects whose form inhabits a non-Euclidian space. Pouteau analyzes how the open character of plants is modified in the process of domestication, and examines various ways in which moral concern towards plants can be expressed. She further suggests that interdisciplinary art–science–philosophical research could play a vital role in enabling more participatory approaches to the design of plant–human interactions in agriculture and landscape management. Again the suggestion is offered that plant ethics harbors philosophical potential that goes beyond plants themselves, as the notion of an open being can be a cornerstone for developing more general alternative ontologies, epistemologies, and ethical perspectives.

Chapters 8 and 9 explore relational approaches to plant ethics. *Mark Coeckelbergh* questions approaches that fixate on the metaphysics or ontology of the object as unpromising ways to think about an ethics for non-human, non-sentient entities. These approaches are inadequate to plant ethics as much as they are inadequate to other emerging fields, including robot ethics. An alternative, relational approach would proceed in three steps, Coeckelbergh suggests: the exploration of the first-person (human) perspective; the opening to a second-person perspective that takes in the needs of plants (or robots); and the establishment of an active relation to the entity from which an ethics of practice can emerge. In this way, questions about how particular plants, robots, and other non-human, non-sentient entities should be treated turn into questions about the skills and know-how necessary to a good performance of the relational practices through which humans engage these entities.

Maria Schörgenhumer explores what caring for plants means in a relational perspective, and takes gardening as the model practice of plants care. She argues that care, sometimes framed as a virtue (and often described as a virtue of gardening in particular), should rather be understood as a relational mode that demands specific virtues. Her chapter analyzes the relation of care established between persons and plants, and highlights its evaluative and normative aspects, which turn out to be closely linked to an aesthetic account of plant flourishing. It further explains how care can fail, and which virtues are required for good plants care.

Chapter 10 inaugurates the second part of the volume, which is dedicated to appreciations of plant life and applications of plant ethics to real-life issues. The first contribution is by *Robin Attfield*, who delineates the field of forest ethics, its scope, the justification for studying it, and the kind of arguments that are salient to it. Attfield notes that forest ethics is in some respects a branch of plant ethics but also includes topics that lie beyond that field, such as the ethics of forestry. He also vindicates biocentrism as preferable to other value-systems when it comes to plant and forest ethics, and explains how biocentrism articulates the central insight that human needs, including future human needs, have to be weighed up alongside the interests of trees and other forest creatures, including future trees and forest creatures. Attfield argues that forest ethics incorporates some of the themes of sustainable development and ecological restoration, and gives examples of practices of reforestation and of how some of the causes of deforestation can and should be combated.

Chapters 11 and 12 focus on aesthetic appreciations of nature. These chapters have a general scope but address flora-rich landscapes in particular. *Paolo D'Angelo* explores the relationship between human culture and the agricultural landscape. He reconstructs the evolution of landscape perception, showing how the modern era has gradually downgraded the aesthetics of cultivated landscapes. The typically romantic idea that important and beautiful landscapes are "exceptional", on the one hand, and the prejudice in favor of the "wild" and "natural", on the other hand, have conjured up a marginalization of the aesthetics of all these cultivated landscapes that humans inhabit and transform. D'Angelo, however, also sees a reverse trend at work. He notes how the recognition that all kinds of landscapes deserve appreciation and protection is gradually growing, while the cultivated land is increasingly being perceived as a rare and valuable commodity, fueling aspirations to new – or, rather, old and now re-discovered – lifestyles.

Angelika Krebs explores the aesthetic case for landscape conservation. She explains the concept of a landscape and distinguishes various ways in which we can experience a landscape's "atmosphere". Her central claim is that the experience of beautiful landscapes is not just an enriching option, much less a merely subjective preference, but an essential part of human flourishing. The great and irreplaceable value of beautiful landscapes is that they can make us feel at home in the world. This aesthetic argument provides a metaphysically innocent understanding of our feeling that we are part of nature and should try to fit in rather than stand out of it. It also explains, in a thoroughly humanistic mood, why most of us feel a taste of sadness, indignation, and sometimes even horror when witnessing the loss of natural landscapes to human development and activities. Many, if not all, of these landscapes include flora.

But it is not just beauty that is lost when human developments and activities swallow up nature and its plants: we also lose what *Paul Thompson*, in his contribution, calls "cultural services". While mainstream environmental risk assessment recognizes a large category of potential hazards that can be associated with the cultivation of plants – including impacts on biodiversity, food safety, and unintended impacts on water and microclimates – more recent literature on ecosystem services acknowledges that both cultivated and uncultivated ecosystems can be crucial for generating or providing cultural goods. These include a sense of place, cultural identity, and social capital associated with community solidarity. Thompson argues that accommodating the cultural significance of plants within a risk assessment and public engagement framework may require scientists and policy makers to revise their views on the ontology of plants themselves.

Chapter 14 and 15 focus on gardens. *Christian Dürnberger* argues that utopian thinking, often thought to be bankrupt or even dead these days, is in fact very much alive particularly with respect to ecological questions. Utopian as well as dystopian content can be found in current disputes on the human–nature relation, agriculture, and food. What is new is that utopian thinking seems to be less focused on public policy and the design of ideal societies, and more on the life-style of individuals. One way of describing the latter trend is to say that utopian thinking withdrew from public places and moved into the garden: to the definition of manageable practices that might give individuals a sense of meaining in the face of vast ecological changes and challenges and of the vast, globalized, systemically interconnected human systems that have brought them about.

David Cooper explores this search for meaning in and through gardening. He focuses on food gardening and asks whether and how "growing your own" can contribute to the good life. Cooper argues that good gardening is a practice that calls upon, cultivates, and exercises various virtues. In doing so, it invokes and manifests a live, practical understanding of important truths about the living world and the relation of that world to human culture. To that extent, Cooper concludes, good gardeners are both virtuous and "in the truth". These are excellent personal reasons for people to "grow their own".

The last four chapters of the volume deal with novel human technologies involving plants in various ways. *Nicole Karafyllis* discusses the complex philosophical and normative implications of seed banks – organizations that store and preserve plant seeds, especially of varieties that are rare, have fallen out of commercial use, or may have unique desirable genetic characteristics. This raises questions about whether a seed in long-term storage should be regarded as a plant, about the nexus of plant generation and proliferation that is interrupted by seed banking in a way that compromises plants' "wildness", and about the normative aims of seed banking, which may range from banking future breeding options to conserving the "common heritage of mankind". Karafyllis argues that reflecting philosophically on seed banking may lead us to challenge approaches to plant ethics that take "wild(er)ness" as a criterion of value, or that assume individual plants to be the one unit of moral concern. She claims that even plants that are shaped by humans into "bio-facts" – techno-natural entities that respect no distinctions between the natural and the artificial, as are the plants grown from banked seeds – deserve ethical consideration.

Frauke Pirscher discusses CRISPR/Cas in crop breeding: a genome editing technique that transfers genes mainly within species boundaries. The resulting plants can be identical to the outcomes of conventional breeding techniques, or even to their wild relatives. This triggers societal debates over the acceptability of CRISPR/Cas in organic farming, and more generally over whether the plants obtained through CRISPR/Cas should or should not count as genetically modified, particularly for the purposes of regulation. Pirscher analyzes how different concepts of naturalness lead to different moral judgments about the moral acceptability of CRISPR/Cas techniques. She argues that even though plants obtained through CRISPR/Cas techniques may be indistinguishable from those obtained through traditional breeding techniques or even their wild relatives, this indistinguishibility of outcomes is not enough to make CRISPR/Cas ethically uncontroversial.

Bart Gremmen and *Vincent Blok* discuss another momentous innovation in plants manipulation and cultivation, namely precision agriculture. Precision agriculture technologies can be conceived as creating "digital" plants by integrating huge amounts of data of various kinds. At first sight, this looks like the next step towards the total instrumentalization of crop plants. Using Kallhoff's notion of plant flourishing as an evaluative yardstick, however, Blok and Gremmen scrutinize from the perspective of plant ethics the acceptability of various technologies that are key to precision farming, including sensing technologies, data processing and utilization, and control technologies. The authors argue that "responsible" precision farming can satisfy important requirements of a flourishing-based plant ethics, and individuate some important criteria for qualifying precision farming as "responsible".

The volume closes with what is perhaps the most futuristic application of plant studies: the case of the Plantoid, a robot replica of an organic plant root that mimics root adaptive behaviors and strategies for soil exploration. Its inventors, *Barbara Mazzolai* and *Pericle Salvini*, explain how the artificial roots are able to grow and penetrate the soil, and how the robot "perceives" and responds to its environment. They then go on to discuss the ethical implications of robotic plants. Could plant robots, for example, destabilize the relationship between humans and plants in significant ways? If so, would that necessarily be for the worse? For centuries, plants have been thought to be inferior to other living entities, incapable of moving, without intelligence and sensorial and perceiving capacities. Mazzolai and Salvini suggest that plantoids could in fact improve human relations to plants, by raising awareness of plants' complex behaviors, adaptive strategies, and unobvious and not yet fully understood capacities.

These 19 contributions make up for the first volume thematically dedicated to plant ethics. They paint an intentionally mosaic picture whose aim is to give readers a sense of just how complex and multi-faceted is the role of plants in our lives. The book is not meant to set any truth in stone: it is a seed that we plant in the field of environmental ethics, in other fields of applied ethics, and of philosophy more generally. We hope that it will branch out to expand our philosophical imagination in new directions.

References

Attfield, R. 1981. The good of trees. *Journal of Value Inquiry* 15, no. 1: 35–54.

Attfield, R. 1991. *The ethics of environmental concern*. Athens and London: University of Georgia Press.

Hall, M. 2011. *Plants as persons: A philosophical botany*. Albany, NY: State University of New York Press.

Ingensiep, H., W. 2001. *Geschichte der Pflanzenseele. Philosophische und biologische Entwürfe von der Antike bis zur* Gegen*wart*. Stuttgart: Alfred Kröner Verlag.

Kallhoff, A. 2002. *Prinzipien der Pflanzenethik. Die Bewertung pflanzlichen Lebens in Biologie und Philosophie*. Frankfurt; New York: Campus.

Mabey, R. 2015. *The cabaret of plants: Botany and the imagination*. London: Profile Books.

Mancuso, S. and Viola, A. 2015. *Brilliant green. The surprising history and science of plant intelligence*. Translated by Joan Benham. Washington; Covelo; London: Island Press.

Marder, M. 2013. *Plant-thinking: A philosophy of vegetal life*. New York: Columbia University Press.

Stone, C., D. 1996. *Should trees have standing? And other essays on law, moral and the environment*. Oxford: Oxford University Press.

Part I

Concepts and approaches

Setting the stage for plant ethics

1 The value of plants

On the axiologies of plants

Gianfranco Pellegrino

Introduction

Plant ethics is the latest attempt at extending the scope of moral considerability. Ethics began with the nearest and dearest, came to include members of the same tribe, then co-nationals, then human beings at large, then animals, and now finally plants (see Kallhoff 2002; Pouteau 2014). Here, I focus on various ways to include plants in the realm of morally valuable things – that is, on various *axiologies of plants*.

For some authors, moral value can be grounded in *aesthetic* value.[1] Here, I do not pursue this path, because I explore the idea that plants have value *independently of valuers* – whereas aesthetic value is very often construed as a function of aesthetic appreciation.[2] In what follows, I assume that value grounds *pro tanto* reasons to bring about or protect the existence of valuable things.

First, I illustrate some relevant intuitions and background distinctions. In the following section, I present some plant axiologies – emphasizing how one strategy is common to all of them: they all *extend* to plants the grounds of the value of *non-plants*.[3] I then list three challenges to this strategy before I finally propose an alternative strategy in the last section of the chapter.

Intuitions and distinctions

This section articulates intuitions about the value of plants, as well as some relevant distinctions framed in the language of value theory.

Intuitions

Mary Midgley imagines Robinson Crusoe wondering whether he should "devastate" his island after his leave, by "settling sparks and powder craftily among certain dry spinneys" chosen for that purpose. Midgley suggests that Crusoe would feel "an invincible objection to this senseless destruction". So do the rest of us, she claims. Unfortunately, Midgley contends, the actual content of this objection is difficult to state, at least within "the language of our moral tradition" (Midgley 1983, 36). This tradition, whose main representative for

Midgley is contractarianism broadly understood, is blind to many potential bearers of value (cf. also Nussbaum 2006).[4]

Crusoe's island hosted animals, plants, and other inanimate natural items (putting aside Crusoe himself, Friday, and his people). Midgley's question may be stated as follows: Are non-human beings – such as non-human animals, plants, inanimate elements of nature, and eco-systems – valuable?[5] If so, are they valuable merely as means for human purposes and objects of human concerns? Or do they have independent value, that is, are they valuable independently of human ends, stances, and considerations? Midgley's contention is that most of us think that many non-human beings have independent value. Call this the Independent Value Intuition, or Independent Value, for short (cf. Agar 2001, 83; Varner 1998, 80, 95). In this chapter, I focus on a narrower version of Midgley's question, to be put as follows: Are plants valuable?[6] If so, is their value dependent on human or animal ends and concerns, or is it independent of them, or both?

To this narrower question, one may answer in the affirmative. Compare a planet without life, "Lifeless", with a planet with no animal and human life but replete with plant life, "Flora". Assume that both planets are too distant to be reached by any human, super-human or animal being. This excludes the possibility that the two planets be in any way useful or otherwise of concern to humans. Nonetheless, if asked to choose a planet to destroy, most of us would likely have Lifeless destroyed and save Flora. This judgment may show that Flora has value, at least as compared to Lifeless, and that this value is independent from its usefulness or observers' concerns (see Scherer 1983).[7] This judgment expresses the intuition that plants have Independent Value.[8] The main objective of the following sections is to present various philosophical accounts of this intuition, that is various plant axiologies.

Distinctions[9]

Any discussion of value may refer to the following topics:

1. The *source* of value, namely what or who establishes that things are valuable. On *subjective views*, *valuers* are the source of value – things are good or bad because they are regarded as such by beings able to entertain, and often to express, value judgments. Often, such beings are identified with human beings. Accordingly, nothing can have value if there are no humans, or if human presence and concerns are not assumed to be at play. In a subjective plant axiology, plants have value because human beings regard them as having value. Plants cannot have value if there are no human beings around, or if human judgments are not assumed to be at play. On these views, the value of plants *depends* on human valuing.

   On *objective views*, things have value independently of valuers and their judgments. In an objective plant axiology, plants can have value even in a world without human beings. Independent Value amounts here to the idea that plants' value is *objective*: *independent of human valuers and valuing.*[10]

2. The *attitude* to value. Suppose I claim that kindness is good. If asked: "Why is kindness good?" I can mention a further valuable thing to which kindness conduces – "Kindness yields pleasure, and pleasure is good". Of course, the same question can then be raised concerning pleasure. To this, I can answer: "Well, it is obvious that pleasure is good – it is *good in itself, as an end*". These two answers distinguish between means to pleasure and the pleasure that is obtained by these means – that is, pleasure regarded as an *end*. Both things may be valued – but the former are valued as *means*, and the latter are valued as ends, *for their own sake*, or *in themselves*. Independent Value in this sense is the idea that plants are not merely, and not always, instruments for human ends.[11] They can be given value independently of human concerns and aims – as ends in themselves, for their own sake: they can be given *final* value. The independent value of plants is a value *independent of human aims*.
3. The *grounds* of value – i.e. the features of a thing that make it valuable, its *value-making features*. Kindness is valuable as a means because it leads to pleasure. Its conduciveness to pleasure is a value-making feature of kindness. Pleasure is valuable as an end because of some of its features – for instance the fact that pleasant experiences are evolutionarily adaptive. The value-making features can belong to the intrinsic nature of the valuable thing – they can be *intrinsic* properties of it – or they can be *extrinsic* features – i.e. properties that the valuable thing has as a *relatum*, or a part and so on. When something has value *because* of its intrinsic properties, then it has *intrinsic* value. By contrast, when the value of something depends on its extrinsic or relational properties, then that something has *extrinsic* value.[12]

   In this case, Independent Value means intrinsic value. The independent value of plants is the value that plants have independently of any relation they do or can entertain with other (valuable and not valuable) things.

In what follows, I explore strategies employed to identify the value-making features of plants.

Plant axiologies: The extensionist strategy

Intrinsic value is grounded in the intrinsic properties of valuable things. Accordingly, any substantive axiology amounts to a view regarding which intrinsic properties ground intrinsic value. As a consequence, any plant axiology should single out features of plants able to ground value: any plant axiology depends on philosophical views of the nature of plants – on a philosophical botany (see Hall 2011, 4; Pouteau 2014, 23).

A plant axiology can be obtained by *extension*. An extensionist strategy rests on the following reasoning: Human beings and/or non-human animals have value because they have sentience, or because they have autonomy, or agency, or any other feature generally assumed to ground intrinsic value. As plants can share sentience, or autonomy, or agency, or some of the other grounds of intrinsic value, then they have intrinsic value as well.[13]

The main extensionist plant axiologies, ranging from ecocentric and biocentric approaches to concepts of plant intelligence, are the following:

1. *Constitutive value of plants: plants as parts of ecosystems (ecocentrism)*. According to some authors, ecosystems have value as such. Assuming that the presence of plants is necessary to the proper constitution and workings of ecosystems, the value that plants contribute to the whole is a form of intrinsic value – it is constitutive of the value of the whole (see Attfield 1981; Johnson 1991, chaps. 3 and 4; Leopold 1949; Rolston 1988, chap. 5, 2002).
2. *Plants as living beings (biocentrism)*. *Being alive* is a value-making feature – as life itself is a valuable thing. Plants are individual living things (in virtue of their tendency to growth and resilience). If so, they have value, like many other living beings.[14]
3. *Sentience, or plants as animals (zoocentrism)*. Some authors view the capacity of feeling, and in particular of feeling pleasures and pains, as grounds for value.[15] Since Darwin, the study of plant signalling has emphasized plants' capacity to move purposefully and to feel. Plants, then, are to be seen as active sentient beings, capable of complex interactions with their environments. In virtue of their sentience, plants have intrinsic value (see Hall 2011, 38–43, 156; Pouteau 2014, 7–8).
4. *Flourishing, interests, integrity, dignity, and legal rights (perfectionism)*. For some authors, plants are organisms with a typical life-cycle and a typical growth process, consisting of a lifetime development that includes the realization of typical characteristics. Accordingly, plants have a good of their own – they can flourish. Even though plants can lack mentality – or can have a merely non-subjective form of intelligence – the life of plants can go well or badly, and their going well is valuable – as it is valuable that human lives go well.[16]

   Some other writers claim that plants have value in so far as they have interests, and that they can have interests in so far as they are subjects of a life, and have a good of their kind (see Nussbaum 2006; Rolston 1988, 108).

   According to others yet, plants have the necessary standing to be granted *legal rights* – that is to have legal suits pursued in their own name (and not of some humans that have a stake in them), where damages are calculated in terms of loss to plants and judgment is applied to their benefit.[17] One typical way of pressing this idea is by assimilating plants – understood as inanimate objects – to *legal persons* such as trusts, corporations, joint ventures, nation states, and so on, or to *legal incompetents* – for instance humans in vegetative states – for whom a guardian can be appointed. This view may rest on the idea that a right-holder is also a valuable thing – a controversial idea, which I do not discuss here (see Kantor 1980; Norton 1982; Wellman 1998, 122–4).

   Finally, some authors have claimed that plants are autonomous and self-ordering wholes, with a capacity to reach certain natural aims. This feature has been called the *integrity* of plants, and it has been taken as grounds of their value (see Bueren and Struik 2005).

5. *Intelligence and personhood*. Plants are "other-than-human persons", sharing with human beings "many capacities and capabilities" (Hall 2011, 161, 155). Plants have a form of active intelligence, consisting in their capacity to adapt to their environments by growing in certain ways and not others, processing information, engage in basic decision making, problem solving and communication, as well as some form of autonomous choice (see Baluska et al. 2006; Brenner et al. 2006; Hall 2011, 158, 160; Ryan 2012; Trewavas 2003, 2005). In virtue of their being intelligent, autonomous, and subjects of a personal life, like animals and human beings plants can be ascribed with dignity or worth.[18] As a consequence, they have value, and this value is independent from human attitudes and concerns (see also Hallé 2002, 302; Marder 2013).

All the positions described above are instances of the extensionist strategy. However, the extensionist strategy is not free of problems.

Three challenges to the extensionist strategy

The extensionist strategy faces the following three objections:

A. *The descriptive objection*. In most cases the extension looks strained. Descriptions of plants as living beings, sentient creatures, or intelligent non-human persons may be vague as well as inaccurate, as plants' life, sentience or intelligence are utterly different from animal or human life, sentience or intelligence. (Not to mention the difficulty of giving uncontroversial definitions of life, sentience, or intelligence even in non-human animals or humans.) Consider, for example, individuality. Plants' life is not the typical life of distinct individuals, as is the case with humans or non-human animals. Rather, it is mostly the life of aggregates. Take fruit trees: *qua* individuals, do they include their fruits? And what about these fruits when detached from the trees? In a sense, life continues in the fruits – they change their shapes in various ways in virtue of various chemical reactions, and these changes follow a sort of life-cycle – and then, arguably, the fruits "die" when they become rotten. Was this all one individual, or an aggregation of many individual things of different kinds? Take, next, plant sentience and intelligence. Plants neurobiology suggests that plants "may actually have thousands of brain-like entities", rather than a "centralized brain and/or nervous system" (Hall 2011, 147).[19] Plants may surely be living, sentient, and intelligent beings – but they are so in their own, specific ways, which differ radically from the ways in which humans or animals are and which make them valuable.

   If what plants possess are rather distant counterparts of animal or human value-making features, simply extending value from animals or humans to plants may be inappropriate. For example, the fact the human intelligence is valuable (or it is a value-making feature of human beings) does not warrant the conclusion that vegetal intelligence is valuable, too – or that it is valuable for the same reasons and in the same ways in which human intelligence is. Thus, since plants' features are distant counterparts of non-plants' value-making features, the extension of value from non-plants to plants is not necessarily warranted.

B. *The receptacles of value objection.* In the extensionist strategy, the value of plants is simply *derived* from the value of non-plants. Plants are simply *carriers*, *vessels* or *receptacles* of non-vegetal values. It is not the case that they have value *qua* plants.

This objection rests on a distinction between intrinsic and *essential* properties. Life, sentience, and intelligence can be intrinsic properties of plants – that is properties that plants have in virtue of the beings that they are – but they may not be *essential* properties of plants – that is properties that identify plants as such, in any possible world where plants exist.[20] Conceivably, plants can recognizably exist without sentience or intelligence, or with relevantly different kinds of sentience or intelligence, and still be correctly regarded as plants.[21] As a consequence, the extensionist strategy can enlighten the intrinsic value of plants but it is unable to provide an account of plants' *essential* value, that is of their value *as* plants, as beings different both from animals and humans (see Pouteau 2014). Yet, that is arguably the value that plants should have if the idea that they have Independent Value is to be vindicated. For Independent Value is the thesis that plants have value *qua* plants. Consider a comparison of three planets, Flora, Lifeless, and Bacteria, the latter being a world populated only by bacteria. A plausible view might be that it is better to preserve Flora than to preserve Bacteria, and better to preserve Bacteria than to preserve Lifeless. If this is a plausible verdict, than plants have greater value than mere living things, and this value should be grounded in features different from being alive. These features may be the essential properties of plants: the properties that make them plants.[22]

C. *The inherent value objection.* In viewing plants as bare receptacles of general value, the extensionist strategy cannot account for the value of *particular* plants – or for the particular value of plants. Consider the well-known debate about plastic trees and the many discussions about the preservation of historical or monumental trees (see Sagoff 1974; Tribe 1974). It seems that in these cases particular plants have value as such, as the particular specimens that they are. Some of these plants are viewed as valuable for their age, or for their rarity, and they are often considered irreplaceable.[23] These specimens have value not on account of instantiating general valuable properties but because of the particulars that they are. It is not the case that a century-old redwood has value primarily because of its sentience, or intelligence, or long life – nor does it have value primarily as a specimen of its species. It is the specific, particular plant that it is that has value. A century-old redwood has *inherent value* – that is it has value as a valuable individual, not as a mere receptacle, or instantiation, of general value (see Regan 2003, 67). The extentionist strategies listed above cannot account for the particular, or inherent, value of plants.[24]

Plants as particular givens: A non-extensionist plant axiology

Objections B and C above can be answered by giving an account of the particular value of plants – an account that will be non-extensionist, because it will not derive the value of plants from the value-making properties of non-plants. Such non-extensionist strategy may also escape objection A, as it will not rely on strained extensions of value-making properties from and to radically different entities. Giving a complete account of such a view is impossible within the confines of this chapter, but I will try to sketch it.

Ronald Dworkin has distinguished between two kinds of intrinsically valuable things. Some things are *incrementally valuable* – the more of them we have the better – others are *inviolable* or *sacred*. For example knowledge is incrementally valuable: we want to know as much as we can, so we always want to *bring about* more knowledge. Non-incrementally and inviolable valuable items, on the other hand, are typically items that we do not want to bring about but that we value once they already exist. Dworkin mentions examples of inviolable value – human life, works of art, and endangered natural species:

> We think it very important, and worth considerable economic expense, to protect endangered species from destruction at human hands or by a human enterprise. [. . .] We consider it a kind of cosmic shame when a species that nature has developed ceases, through human actions, to exist.
>
> (Dworkin 1993, 75)

Applied to natural species, Dworkin's view ascribes value to their *genesis*, namely to thc unique and humanly irreproducible process through which nature created them. Dworkin views the "evolution of species" as "a process of creation" (see also Hargrove 1988, chap. 6). The destruction of unique species would be a "waste of nature's investment", meaning "not that nature is a conscious investor but that even unconscious natural processes of creation should be treated as investments worthy of respect" (Dworkin 1993, 79). We find natural items valuable in this specific way because they have been produced by long, and remarkable, natural processes:

> We honor nature because it has produced striking geological formations and majestic plants and living creatures we find extraordinary, including us, and we protect examples of that production—mountains or rivers or forests or animals—in a special and more intense way because they are natural.
>
> (Dworkin 1993, 80–1)

Dworkin's account has many complexities, and it should be further elaborated before applying it to plants (see O'Day 1999). However, Dworkin's is already an improvement on previous similar accounts. Consider two among the most influential.

Robert Elliot (1997) claimed that wild nature's value is grounded in the specific property of *naturalness* – to be understood as both the "property of being naturally evolved" and the property of being separate, radically different from the human and artificial world of culture and technology, of being raw and unshaped or unmarked by human intentions and design. For Elliot, this radically non-artifactual nature of wilderness grounds its intrinsic value: the complexity and independence of natural processes elicit positive value attitudes and responses in human observers (see Elliot 1997, 59–60; see also Brennan 1984; Godlovitch 1994; Williams 1995).

Robert Goodin (1992) suggests that, in a green theory of value, the value of natural items is grounded in "the history and process of their creation", and particularly in the fact that they "have a history of having been created by natural processes rather than artificial human ones". This theory of value is both historical and process-based. Moreover, according to Goodin there is a deeper feature grounding the value of naturalness – understood as being the outcome of natural processes. These natural processes provide context to, and make sense of, human lives – a context that is independent, larger than, and "outside of ourselves", and which cannot be reproduced by human forgery without losing its sense-making power. The value of nature, according to Goodin, is grounded in its working as a necessary larger context to human lives. Nature satisfies the deep psychological need of people to locate themselves within a wider, sense-making context (see Goodin 1992, 26–7, 37–9, 42–54).[25]

To call a given feature good in itself, or good as an end, is (or should be) where all questions of value stop. If a given feature is a plausible final end, then no further questions on its value should be necessary. The same holds when a given feature is selected as grounds of intrinsic value. Pleasure is obviously good, and its goodness grounds and explains the intrinsic value of many pleasant experiences. But neither Elliot nor Goodin indicate a plausible point at which questions about the intrinsic value of nature and naturalness should stop. It still seems possible to ask them: "Why does being the outcome of natural evolution, or being part of a larger independent context, ground the value of natural items?"

Dworkin is pointing to a different value-making feature than Elliott or Goodin. His view is that natural items are the outcome of a sort of investment of nature, partially similar to human investments: natural items are an achievement or work of nature. Investments make their outcomes valuable in so far as these outcomes can be described as achievements. Even an only partially successful effort is often more valuable than a merely accidental success. A successful effort is obviously valuable, at least *prima facie*.[26] Dworkin's suggestion is that nature's workings in natural selection have the same value-making power as human investments (Dworkin notices that we have a tendency to "describe animal species as not just accidents but as achievements of adaptation, as something that nature has not just produced but wrought", Dworkin (1993, 76). Nature's operation in evolutionary history is a form of investment, adding value to its final outcomes.[27]

However, Dworkin's account is still an extensionist strategy, in which particular specimens of plants are valuable only in so far as they instantiate a general valuable property – being the outcome of nature's investment. This does not rescue

Dworkin's view from objections B and C: individual specimens of nature, and specifically plants, still seem not having value as individuals but only as vessels or receptacles of value.

A better move is made by G. A. Cohen (2011), who claims that certain particular things have value as particular bearers of intrinsic value. Cohen writes:

> If an existing thing has intrinsic value, then we have reason to regret its destruction *as such*, a reason that we would not have if we cared only about the value that the thing carries or instantiates. [. . .] If you happily replace a fine statue by a merely somewhat better one, the production of which requires destruction of the original statue, then you mistreat the now destroyed work as (so to speak) having had the merely instrumental value of being a vessel of aesthetic value.
>
> (Cohen 2011, 210)

For Cohen, the point is that "particular *bearers* of value" should be valued as such – and not simply as vessels of general value: general value, Cohen writes provocatively, "is not the only thing that is valuable: so are particular valuable things" (Cohen 2011, 212). This mode of valuation leads to a "propensity" to "conserve, to not destroy and, therefore, to not replace, even (within limits) by something more valuable [. . .] whatever merely *happens* to exist" (Cohen 2011, 213, 206). The fact that something of value is already there is a reason to keep it, to preserve it: this is the *preservation "of the given* (as such)" (Cohen 2011, 209–10). In this way Cohen identifies a distinctive kind of intrinsic value – *particular* value, that is the value of particular individuals – which must be distinguished from the general value that particular things can instantiate (see Brennan and Hamlin 2016, 345).

The value of particular plants – of century-old redwoods, for instance – may be construed by putting together suggestions from Dworkin and Cohen. In a Cohenite–Dworkinian account of the value of plants, specimens of plants are valuable as particulars: as individual manifestations of processes that bring about the manifold variety of plants. Individual plant specimens have *particular value*: they are valuable as existing, visible and infinitely differentiated outcomes of the investment of nature, of the millennial evolution and adaptation of natural life on Earth. This process produced not only species but all the individual plants that compose these species, and these plants have a value as individuals, over and above their value as members of their species. Their individuality, and their historicity, has value. This specificity is what gives them value as individuals.[28]

Four consequences of this view need to be made explicit. First, and contrary to Dworkin's first quote above, individual plants have greater value than their species. Specific tokens have a value that is not possessed by just any tokens that represents the relevant type (cf. Brennan and Hamlin 2016, 345). On this view, the value of rarity can be accounted for, but not only nor primarily in terms of species survival: the last specimen left of a dying species is immensely valuable, but not only nor primarily because it guarantees the continuance of the natural kind it belongs to. Rather, its value is also, and primarily, grounded in its being a unique individual, whose constitution, shape, pedigree, and appearance is uniquely placed in the natural history of Earth. The last redwood is not valuable only or mainly as the only survived specimen of *Sequoia sempervirens*, but above all as an individually remarkable vestige of the long evolutionary history of the natural selection and spreading of redwoods on Earth.

Second, and related, the ground of value of individual natural items is at least partially coincident with that of individual works of art and other artificial items. The last redwood and the last Greek temple in Sicily are both valuable in virtue of their rarity and of their history. However, the redwood and the temple can be distinguished by looking at their origin and provenance. The histories that make them valuable are different: the temple is one culmination of the history of architecture (a fully human product), the redwood is one culmination of the history of tree selection and propagation on Earth (a predominantly natural series of events) (cf. Ashby 1980, 28–9).

Third, these differences are not all-or-nothing. The history of architecture is partly determined by nature – at least because it is constrained by nature. The natural selection and spreading of redwoods is partly determined by human activities – at least insofar as human beings favoured or at least did not inhibit it.

Fourth, natural and human history can generate value in tandem, adding to one another. An individual plant resulting from both natural selection and human cultivation, for example, has greater value than a wild plant or a piece of art, because it receives value from both human and natural investments.

The Cohenite–Dworkinian view of the value of plants presented above needs refinement, and the implications that it supports need further defense. In particular, principles should be framed to establish whether and when the particular value of plants can override the general value of species and other valuable things, such as animals, persons, and so on. In addition, the difference and similarities between natural and human investments need to be further specified.

However, even in this sketchy form, this view is a better approximation to Independent Value, at least as applied to plants, than any extensionist account. Plants are specific entities – and their specificity is embedded not in species, but in particular items. Plants are valuable as particular givens.[29]

Notes

1. See Brady (1998, 2013, chap. 8), Brady et al. (1999, chap. 10), Carlson (1984, 2010), Godlovitch (1989, 1994), Lynch (1996) and Rolston (1998).
2. See Crisp (1994), Elliot (1997, 58, 61–73), Hargrove (1988, chap. 6, 1992), Hartshorne (1979, 55) and Rolston (1988, 233).
3. The neurobiology of plants suggests that the differences between plants and animals are blurred; see Hallé (2002) and Trewavas (2003, 2005). Here, I assume that a dichotomy between plants and non-plants can be kept without positing a neat distinction. Whatever plants are, non-plants are simply entities not strictly identical with plants.

4. Midgley does not engage in value talk; she puts the issue in terms of rights and duties (see Midgley 1983, 40–2). Here, I take these two ways of talking as equivalent, at least because I assume that value grounds reasons to act.
5. In environmental ethics, views giving value to individuals are distinguished from views ascribing value to wholes – that is to eco-systems or species. Animal rights or animal sentience theories are a main sample of individualism; see Regan (2003) and Singer (1975). Biocentrism or ecocentrism are samples of holism; see Goodpaster (1978) and Johnson (1991).
6. Vascular or higher plants are commonly distinguished from non-vascular plants. Some of the views I consider below are restricted only to vascular plants, while broadly biocentric views can be extended also to non-vascular plants and other micro-organisms. See Hallé (2002, p. 18) and Kallhoff (2014, 2 n. 2).
7. This thought experiment is an instance of the well-known isolation test for value-bearers – see Moore (1993a, 135) – as well as "last man cases": see Attfield (1983, 155), Benn (1977), Crisp (1994, 77–80), Elliot (1997, 55, 62–9), Goodin (1992, 43, 83), Grey (2009), Hargrove (1988, 171–4), Lemos (1994, 10–11), Moore (1993a, 236), Routley (1973), Routley and Routley (1989) and Singer (1979).
8. I assume that Flora contains *exclusively* vegetal life; I take "animal" life as including micro-organismic and cellular life, such as bacteria or microbic organisms. Hence, the intuition that Flora has value is not the intuition that naked or bare life has value; rather, it is the intuition that *vegetal* life has value.
9. In this section, I rely on Agar (2001, chap. 1), Callicott (1984, 1985, 1995), Crisp (1994), Dancy (2004, 166–70), Dworkin (1993, 71), Elliot (1997); Goodin (1992, chap. 2), Green (1996), Hargrove (1992), Hirose and Olson (2015), Korsgaard (1983), Nussbaum (2006, p. 330), Lemos (1994), Moore (1993b), O'Neill (1992, 1993, chap. 2), O'Neill et al. (2008), Orsi (2015), Pouteau (2014, 3–4); Rabinowicz and Rønnow-Rasmussen (2000), Rolston (1988, chaps. 3 and 6); Schroeder (2016). In environmental ethics, many discussions have been concerned with the contrast between *anthropocentric* versus *non-anthropocentric* value. I avoid this language for reasons effectively explained in Hargrove (1992, esp. 183–4).
10. For subjective views, see Callicott (1989, 133–4) and Hargrove (1992, 196); for objective views, see Crisp (1994), Rolston (1988, 1994, 2002) and Taylor (1986).
11. For instrumental accounts of the value of nature, see Norton (1984), Passmore (1974) and Weston (1985).
12. Here, I assume that a property is intrinsic if something has it in isolation and in virtue of what it is, depending on no other things; see Lewis (1983) and Weatherson and Marshall (2017).
13. On extensionist ethics, see Agar (2001), Darwin (1981, 103), Goodpaster (1978), Hall (2011, 2), Houle (2011), Leopold (1949, 202–4), Nash (1989, 3–12), Nussbaum (2006, 226, 346), Stone (2010, chap. 1) and Wellman (1998). Specifically on plants, see Hall (2011), Kallhoff (2014) and Pouteau (2014).
14. See Agar (2001, chaps. 4–6), Goodpaster (1978), Rolston (1988, chap. 3, 1994), Sterba (1995), Taylor (1986) and Varner (1998). On resilience, see Bueren and Struik (2005, 483).
15. See Donaldson and Kymlicka (2011, chap. 2), Francione (2000, 2006), Frankena (1979), Singer (1976), Steiner (2008), Stone (2010, 189 n. 92) and Warnock (1971, 151).
16. See Attfield (1981), Feinberg (1974, 51), Goodpaster (1978, 319), Kallhoff (2014), Rolston (1988, chap. 3) and Taylor (1986, 45, 66). Cp. Crisp (1994, 81) and Wellman (1998, 118–9).
17. Extending to plants the view ascribing moral rights of nature is much more controversial, though; see Donaldson and Kymlicka (2011, 32–6); Hartshorne (1979), Nash (1989), Povilitis (1980) and Wellman (1998, 118–22).

18. The 1992 Swiss constitution established that all living creatures – plants included – have dignity; see Brom (2000), Koechlin (2009) and Pouteau (2014, 11–13).
19. See Alpi et al. (2007), Firn (2004), Struik et al. (2008) and Trewavas (2004). Cp. also Agar (2001, 29–30, 56–7, 59–61), Hallé (2002, 108–24) and Pouteau (2014, 21–2).
20. The definition of essential properties is notoriously complicated. Cp. Robertson and Atkins (2016).
21. What about life? Consider pine cones. They can be viewed as inactive, or dead, tissues. But they can move, and they are obviously plants, or parts of plants – fruits. If so, they are plants even though they are not alive. Then, being alive is not an essential feature of plants. Also, being alive is not a value-making feature, as it might be argued that pine cones, or other fruits, can have independent value, *qua* plants. If Flora is replete with pine cones, and it is compared with Lifeless, the intuition still would be that Flora has value. Cp. Dawson et al. (1997).
22. It might be argued that a planet populated only by plants, without bacteria and microbial organisms, is scientifically impossible. Be this as it may, the mere logical possibility of such a scenario is enough to our purposes here.
23. However, irreplaceability is not necessarily a feature of valuable historic items; see Matthes (2013). See also Goodin (1992, 57–61).
24. Objections B and C have been commonly levelled against utilitarianism; see Regan (2003, 59) and Singer (1993, 121). See also Wilson (2006).
25. On the value of historic items, see Matthes (2013, 2016). On history as a dimension of naturalness, see Siipi (2008).
26. Mischievous efforts are bad, of course. But it seems plausible to say that they are less bad as efforts than they would be as accidents. Hitler's genocidal project is bad, but arguably less bad than a completely fortuitous genocide. I am considering value, not rightness. It might be argued that Hitler's genocidal project is wrong, whereas a completely accidental genocide – following, say, a tsunami hitting a whole people – is bad, but not wrong. I am not suggesting that accidental goods are valueless. Simply, intentionally achieved goods are of greater value than accidental ones.
27. "Final" does not mean "definitive", of course, as evolution can have no end.
28. In being based on history, particular value can be regarded as extrinsic, as it is based on the relation of the valuable things with previous things; cp. O'Day (1999, 207–8). However, this kind of relation is an *internal* one, like the arrangement of parts in a landscape. Historical objects are *essentially constituted* by their relation with the past – and some essential properties are intrinsic. If so, historical value is intrinsic value.
29. Michele Bocchiola and Marcello Di Paola read a previous version of this chapter, giving me precious suggestions. I am grateful to them. Usual *caveats* hold.

References

Agar, N. 2001. *Life's intrinsic value. Science, ethics, and nature*. New York: Columbia University Press.

Alpi, A., Amrhein, N., Bertl, A., Blatt, M., R., Blumwald, E., Cervone, F., Dainty, J., et al. 2007. Plant neurobiology: No brain, no gain? *Trends in Plant Science* 12, no. 4: 135–6.

Ashby, E. 1980. *The search for an environmental ethic. The tanner lectures on human values* (Vol. 1), edited by Sterling M. McMurrin. Salt Lake City, UT: University of Utah Press.

Attfield, R. 1981. The good of trees. *Journal of Value Inquiry* 15, no. 1: 35–54.

Attfield, R. 1983. *The ethics of environmental concern*. New York: Columbia University Press.

Baluska, F., Mancuso, S. and Volkmann, D. eds. 2006. *Communication in plants*. Berlin: Springer-Verlag.

Benn, S. 1977. Personal freedom and environmental ethics: The moral inequality of species. Pp. pp. 401–24 in *Equality and freedom: International and comparative jurisprudence*. Dobbs Ferry: Oceana Publications.

Brady, E. 1998. Imagination and the aesthetic appreciation of nature. *The Journal of Aesthetics and Art Criticism* 56, no. 2: 139–47.

Brady, E. 2013. *The sublime in modern philosophy: Aesthetics, ethics, and nature*. Cambridge: Cambridge University Press.

Brady, E., Jane Howarth, J. and Pratt, V. 1999. *Environment and philosophy*. London: Routledge.

Brennan, A. 1984. The moral standing of natural objects. *Environmental Ethics* 6, no. 1: 35–56.

Brennan, G. and Hamlin, A. 2016. Conservative value. *Monist* 99, no. 4: 352–71.

Brenner, E. D., Stahlberg, R., Mancuso, S., Vivanco, J., Baluška, F. and Volkenburgh Van, E. 2006. Plant neurobiology: An integrated view of plant signaling. *Trends in Plant Science* 11, no. 8: 413–19.

Brom, F. W. A. 2000. The good life of creatures with dignity. Some comments on the Swiss expert opinion. *Journal of Agricultural and Environmental Ethics* 13, no. 1: 53–63.

Bueren, E., T. Lammerts Van and Struik, P., C. 2005. Integrity and rights of plants: Ethical notions in organic plant breeding and propagation. *Journal of Agricultural and Environmental Ethics* 18, no. 5: 479–93.

Callicott, J. B. 1984. Non-anthropocentric value theory and environmental ethics. *American Philosophical Quarterly* 21, no. 4: 299–309.

Callicott, J. B. 1985. Intrinsic value, quantum theory, and environmental ethics. *Environmental Ethics* 7, no. 3: 257–75.

Callicott, J. B. 1989. *In defense of the land ethic: Essays in environmental philosophy*. Albany, NY: State University of New York Press.

Callicott, J. B. 1995. Intrinsic value in nature: A metaethical analysis. *Electronic Journal of Analytical Philosophy* 3, no. 5. http://ejap.louisiana.edu/EJAP/1995.spring/callicott.1995.spring.html.

Carlson, A. 1984. Nature and positive aesthetics. *Environmental Ethics* 6, no. 1: 5–34.

Carlson, A. 2010. Contemporary environmental aesthetics and the requirements of environmentalism. *Environmental Values* 19, no. 3: 289–314.

Cohen, G. A. 2011. Rescuing conservatism: A defense of existing value. Pp. 203–30 in *Reasons and recognition. Essays on the philosophy of T. M. Scanlon*, edited by S. Freeman, R. Kumar and R. J. Wallace. Oxford: Oxford University Press.

Crisp, R. 1994. Values, reasons and the environment. Pp. 75–87 in *Philosophy and the Natural Environment* (Vol. 36). Cambridge: Royal Institute of Philosophy Supplement.

Dancy, J. 2004. *Ethics without principles*. Oxford: Oxford University Press.

Darwin, C. 1981. *The descent of man, and selection in relation to sex*. Princeton, NJ: Princeton University Press.

Dawson, C., Vincent, J. F. V. and Rocca, A. 1997. How pine cones open. *Nature* 390, no. 6661: 37–45.

Donaldson, S. and Kymlicka, W. 2011. *Zoopolis: A political theory of animal rights*. Oxford: Oxford University Press.

Dworkin, R. 1993. *Life's dominion: An argument about abortion, euthanasia, and individual freedom*. New York: Alfred Knopf.

Elliot, R. 1997. *Faking nature: The ethics of environmental restoration*. London and New York: Routledge.

Feinberg, J. 1974. The rights of animals and unborn generations. Pp. 43–68 in *Philosophy and environmental crisis*, edited by W. T. Blackstone. Athens: University of Georgia.

Firn, R. 2004. Plant intelligence: An alternative point of view. *Annals of Botany* 93, no. 4:345–51.

Francione, G. L. 2000. *Introduction to animal rights: Your child or the dog?* Philadelphia, PA: Temple University Press.

Francione, G. L. 2006. Taking sentience seriously. *Journal of Animal Law and Ethics* 1:1–18.

Frankena, W. K. 1979. Ethics and the Environment. Pp. 3–20 in *Ethics and problems of the 21st century*, edited by K. E. Coodpaster and K. M. Sayre. Notre Dame, IN: Indiana University Press.

Godlovitch, S. 1989. Aesthetic protectionism. *Journal of Applied Philosophy* 6, no. 2: 171–80.

Godlovitch, S. 1994. Icebreakers: Environmentalism and natural aesthetics. *Journal of Applied Philosophy* 11, no. 1: 15–30.

Goodin, R., E. 1992. *Green political theory*. Cambridge: Polity Press.

Goodpaster, K., E. 1978. On being morally considerable. *The Journal of Philosophy* 75, no. 6: 308–25.

Green, K. 1996. Two distinctions in environmental goodness. *Environmental Values* 5, no. 1: 31–46.

Grey, W. 2009. Last man arguments. Pp. 40–1 in *The encyclopedia of environmental ethics and philosophy*, edited by J. B. Callicott and R. Robert Frodeman. New York: Macmillan.

Hall, M. 2011. *Plants as persons: A philosophical botany*. New York: State University of New York Press.

Hallé, F. 2002. *In praise of plants*. Portland: Timber Press.

Hargrove, E., C. 1988. *Foundations of environmental ethics*. Englewood Cliffs, NJ: Prentice Hall College Division.

Hargrove, E. C. 1992. Weak anthropocentric intrinsic value. *The Monist* 75, no. 2: 183–207.

Hartshorne, C. 1979. The rights of the subhuman world. *Environmental Ethics* 1, no. 1: 49–60.

Hirose, I. and Olson, J., eds. 2015. *The Oxford handbook of value theory*. New York: Oxford University Press.

Houle, K. 2011. Animal, vegetable, mineral: Ethics as extension or becoming? The case of becoming-plant. *Journal for Critical Animal Studies* IX, no. 1/2: 89–116.

Johnson, L. E. 1991. *A morally deep world. An essay on moral significance and environmental ethics*. Cambridge: Cambridge University Press.

Kallhoff, A. 2002. *Prinzipien der Pflanzenethik: Die Bewertung pflanzlichen Lebens in Biologie und Philosophie*. New York: Campus Verlag.

Kallhoff, A. 2014. Plants in ethics: Why flourishing deserves moral respect. *Environmental Values* 23, no. 6: 685–700.

Kantor, J. E. 1980. The "Interests" of natural objects. *Environmental Ethics* 2, no. 2: 163–71.

Koechlin, F. 2009. The dignity of plants. *Plant Signaling & Behavior* 4, no. 1: 78–9.

Korsgaard, C. M. 1983. Two distinctions in goodness. *The Philosophical Review* 92, no. 2: 169–95.

Lemos, N. M. 1994. *Intrinsic value*. Cambridge: Cambridge University Press.

Leopold, A. 1949. *A sand county almanac, and sketches here and there*. Oxford: Oxford University Press.
Lewis, D. 1983. Extrinsic properties. *Philosophical Studies: An International Journal for Philosophy in the Analytic Tradition* 44, no. 2: 197–200.
Lynch, T. 1996. Deep ecology as an aesthetic movement. *Environmental Values* 5, no. 2: 147–60.
Marder, M. 2013. *Plant-thinking: A philosophy of vegetal life*. New York: Columbia University Press.
Matthes, E. H. 2013. History, value, and irreplaceability. *Ethics* 124, no. 1: 35–64.
Matthes, E. H. 2016. The ethics of historic preservation. *Philosophy Compass* 11, no. 12: 786–94.
Midgley, M. 1983. Duties concerning islands. Of rights and obligations. *Encounter* 60, no. 2: 36–43.
Moore, G., E. 1993a. *Principia ethica*. Cambridge: Cambridge University Press.
Moore, G., E. 1993b. The conception of intrinsic value. In *Principia ethica*, edited by T. Baldwin. Cambridge: Cambridge University Press.
Nash, R. 1989. *The rights of nature: A history of environmental ethics*. Madison, WI: University of Wisconsin Press.
Norton, B. G. 1982. Environmental ethics and nonhuman rights. *Environmental Ethics* 4, no. 1: 17–36.
Norton, B. G. 1984. Environmental ethics and weak anthropocentrism. *Environmental Ethics* 6, no. 2: 131–48.
Nussbaum, M. C. 2006. *Frontiers of justice: Disability, nationality, species membership*. Cambridge, MA: Belknap Press.
O'Day, K. 1999. Intrinsic value and investment. *Utilitas* 11, no. 2: 194–214.
O'Neill, J. 1992. The varieties of intrinsic value. *Monist* 75, no. 2: 119–37.
O'Neill, J. 1993. *Ecology, policy and politics. Human well-being and the natural world*. London: Routledge.
O'Neill, J., Holland, A. and Light, A. 2008. *Environmental values*. London: Routledge.
Orsi, F. 2015. *Value theory*. London: Bloomsbury Academic.
Passmore, J. 1974. *Man's responsibility for nature*. London: Gerald Duckworth & Co Ltd.
Pouteau, S. 2014. Beyond "Second Animals": Making sense of plant ethics. *Journal of Agricultural and Environmental Ethics* 27, no. 1: 1–25.
Povilitis, A. 1980. On assigning rights to animals and nature. *Environmental Ethics* 2, no. 1: 67–71.
Rabinowicz, W. and Rønnow-Rasmussen, T. 2000. A distinction in value: Intrinsic and for its own sake. *Proceedings of the Aristotelian Society* 100, no. 1: 33–51.
Regan, T. 2003. *Animal rights, human wrongs. An introduction to moral philosophy*. Lanham, MD: Rowman & Littlefield Publishers.
Robertson, T. and Atkins, P. 2016. Essential vs. accidental properties. In *The Stanford encyclopedia of philosophy* (Vol. Summer), edited by E. N. Zalta. Stanford, CA: Metaphysics Research Lab, Stanford University. Accessed November 29, 2017 from https://plato.stanford.edu/archives/sum2016/entries/essential-accidental/.
Rolston, H., III. 1988. *Environmental ethics. Duties to and values in the natural world*. Philadelphia, PA: Temple University Press.
Rolston, H., III. 1994. Value in nature and the nature of value. Pp. 13–30 in *Philosophy and the natural environment*, edited by R. Attfield and A. Belsey. Cambridge: Cambridge University Press.

Rolston, H., III. 1998. Aesthetic experience in forests. *Journal of Aesthetics and Art Criticism* 56, no. 2: 157–66.

Rolston, H., III. 2002. What do we mean by the intrinsic value and integrity of plants and animals? Pp. 5–10 in Genetic engineering and the intrinsic value and integrity of animals and plants. *Proceedings of an Ifgene Workshop*, edited by D. Heaf and J. Wirz. Hafan: Ifgene.

Routley, R. 1973. Is there a need for a new, and environmental ethic? Pp. 205–10 in *Proceedings of the XVth World Congress of philosophy*, 17th to 22nd September 1973, Varna, Bulgaria, Varna: Sofia Press.

Routley, R. and Routley, V. 1980. Human chauvinism and environmental ethics. Pp. 96–189 in *Environmental philosophy*, edited by D. S. Mannison, M. A. McRobbie and R. Sylvan. Canberra: Research School of Social Science, Australian National University.

Ryan, J. C. 2012. Passive flora? Reconsidering nature's agency through human–plant studies (HPS). *Societies* 2, no. 3: 101–21.

Sagoff, M. 1974. On preserving the natural environment. *Yale Law Journal* 84, no. 2: 205–67.

Scheffler, S. 1994. *The rejection of consequentialism: A philosophical investigation of the considerations underlying rival moral conceptions*. Oxford: Clarendon Press.

Scherer, D. 1983. Anthropocentrism, atomism, and environmental ethics. Pp. 73–81 in *Ethics and the environment*, edited by D. Scherer and T. Attig. Englewood Cliffs, NJ: Prentice Hall.

Schroeder, M. 2016. Value theory. In *The Stanford encyclopedia of philosophy*, edited by Edward N. Zalta. Stanford, CA: Metaphysics Research Lab, Stanford University. Accessed November 28, 2017 from https://plato.stanford.edu/archives/fall2016/entries/value-theory/.

Siipi, H. 2008. Dimensions of naturalness. *Ethics and the Environment* 13, no. 1: 71–103.

Singer, P. 1975. *Animal liberation: A new ethic for our treatment of animals*. New York: Random House.

Singer, P. 1976. All animals are equal. Pp. 148–62 in *Animal rights and human obligations*, edited by T. Regan and P. Singer. Englewood Cliffs, NJ: Prentice Hall.

Singer, P. 1979. Not for humans only: The place of nonhumans in environmental issues. Pp. 191–206 in *Ethics and problems of the 21st century*, edited by K. E. Coodpaster and K. M. Sayre. Notre Dame, IN: University of Notre Dame Press.

Singer, P. 1993. *Practical ethics*. New York: Cambridge University Press.

Singer, P. 2005. Ethics and intuitions. *Journal of Ethics* 9, no. 3–4: 331–52.

Steiner, G. 2008. *Animals and the moral community: Mental life, moral status, and kinship*. New York: Columbia University Press.

Sterba, J. P. 1995. From biocentric individualism to biocentric pluralism. *Environmental Ethics* 17, no. 2: 191–207.

Stone, D.-C. 2010. *Should trees have standing? Law, morality, and the environment*. Oxford: Oxford University Press.

Struik, P.C., X. Yin and H. Meinke. 2008. Plant neurobiology and green plant intelligence, science, metaphors and nonsense. *Journal of the Science of Food and Agriculure* 88, no. 3: 363–70.

Taylor, P. W. 1986. *Respect for nature: A theory of environmental ethics*. Princeton, NJ: Princeton University Press.

Trewavas, A. 2003. Aspects of plant intelligence. *Annals of Botany* 92, no. 1: 1–20.

Trewavas, A. 2004. Aspects of plant intelligence: An answer to Firn. *Annals of Botany* 93, no. 4: 353–57.

Trewavas, A. 2005. Green plants as intelligent organisms. *Trends in Plant Sciences* 10, no. 9: 413–19.

Tribe, L. H. 1974. Ways not to think about plastic trees: New foundations for environmental law. *Yale Law Journal* 83, no. 7: 1315–46.

Varner, G., E. 1998. *In nature's interests? Interests, animal rights, and environmental ethics.* New York: Oxford University Press.

Warnock, G. J. 1971. *The object of morality*. New York: Methuen.

Weatherson, B. and Marshall, D. 2017. Intrinsic vs. extrinsic properties. In *The Stanford encyclopedia of philosophy* (Fall), edited by Edward N. Zalta. Metaphysics Research Lab, Stanford University. Accessed November 25, 2017 from https://plato.stanford.edu/archives/fall2017/entries/intrinsic-extrinsic/.

Wellman, C. 1998. *The proliferation of rights. Moral progress or empty rhetoric?* Boulder, CO: Westview Press.

Weston, A. 1985. Beyond intrinsic value: Pragmatism in environmental ethics. *Environmental Ethics* 7, no. 4:321–39.

Williams, B. 1995. Must a concern for the environment be centred on human beings? Pp. 233–40 in *Making sense of humanity and other philosophical papers 1982–1993*. Cambridge: Cambridge University Press.

Wilson, S. 2006. Respect of utilitarianism: A response to Regan's "Receptacles of Value" objection. In *Proceedings of the Ohio Philosophical Association* 3. Accessed November 17, 2017 from https://corescholar.libraries.wright.edu/philosophy/2.

2 Utilitarian plant ethics

Tatjana Višak

Introduction

This paper is about utilitarian plant ethics. Ethics, as I see it, concerns the question of what we have reason to do. Utilitarians are welfarists regarding reasons for action. Utilitarians hold that welfare, and only welfare, grounds reasons for action. Utilitarianism is not the only ethical theory that is committed to welfarism. Many other consequentialists are also welfarists. For example, rational egoists hold that only the promotion of one's own welfare gives one reasons for action. Some egalitarians, prioritarians and the like are also welfarists: they hold that only welfare considerations ground reasons for action, and they also have specific ideas regarding how welfare should be distributed.

Utilitarianism is the classical welfarist theory, and it holds in its simplest, act-utilitarian version that one always has most reason to do whatever impartially maximizes welfare. Impartiality here refers to the fact that the welfare of all concerned individuals is taken equally into account: one's own welfare, or that of one's friends, does not count for more than that of others. In contrast to rational egoists, utilitarians are thus rational impartialists.

In light of the nature of utilitarianism, it is clear that our actions concerning plants only relate to what we have reason to do in so far as these actions affect welfare. If plants are welfare subjects, what we do to them matters directly. If they are not welfare subjects, how we treat them matters indirectly, to the extent that it affects welfare subjects. To be clear, a welfare subject is an entity that has a certain level of welfare that can be positive and/or negative and/or neutral.

This paper discusses the question of whether plants are welfare subjects. Before we begin that discussion, however, it is useful to take a brief look at the nature and capacities of plants (Section Two). Utilitarians typically hold that plants are not welfare subjects, and they reach that conclusion on the basis of what they consider to be the correct account of welfare (Section Three). A negative answer to the question of whether plants are welfare subjects, however, can also be obtained without presupposing the truth of any particular account of welfare (Section Four). But even if plants are not welfare subjects, and therefore do not matter directly according to utilitarianism, what we do to plants still matters in so far as it affects the welfare of welfare subjects (Section Five).

Nature and capacities of plants

If asked what plants are, a child would probably point to trees, flowers, grasses, ferns or mosses as common examples of plants. "A plant", according to the Encyclopedia of Life, "is any one of the vast number of organisms within the biological kingdom of Plantae" (Hogan and Taub 2010). The encyclopedia further explains that "these species are considered of limited motility", which means that they lack the ability to move spontaneously and actively (this is not to be confused with mobility, which refers to the ability of an object to be moved). Plants generally make their own food; are characterized by multicellularity and cell structure with walls containing cellulose; and they are capable of photosynthesis (Hogan and Taub 2010).

Plants do not feel pain. When I teach animal ethics and note that a relevant characteristic of pigs, chicken, cows and fish is that they are capable of feeling pain, there tend to be at least several students who remark that plants can also feel pain. These students were instructed at a young age to not tear leaves off the plants in the schoolyard because this hurts the plants. Many among us also routinely speak of plants as if they had feeling, desires, and intentions. One of the best-selling authors in Germany these days is a former ranger who tells his readers, among other things, about the secret lives of trees (Wohlleben 2015). This author and many of his readers literally apply inverted-comma ascriptions of feelings, desires and intentions to plants. I find this philosophically and physiologically incorrect. Plants have *no* nervous system and there is no indication that anything besides a nervous system can lead to conscious experiences, such as pain. Furthermore, pain has the evolutionary function of enabling organisms to avoid tissue damage and the like. Since plants are unable to move away from the threat, pain would not be adaptively useful: it would be useless torture.

Plants can communicate with each other and do so in interesting ways: they release and receive chemicals via the air and their roots, for example. When we humans hear "communication", however, we tend to think of *our* communication – be it written, verbal or visual – and we thus may be tempted to think that communication between plants also happens intentionally as it does among us. However, communication is simply the act of transferring information from one place to another. Non-verbal communication among humans themselves – for instance through body language, gestures or even our scents – often happens unconsciously and unintentionally. Additionally, different organs can communicate with each other in our bodies (Liu and Lin 2017). Machines, too, can communicate with each other. For example, auto racing engineers use machine-to-machine communication to monitor a car's performance. A car's microchip may tell the engine how to operate under various conditions so that the car can achieve the best fuel economy. Neither the microchip nor the engine is conscious, and both lack intentions. According to the Merriam-Webster dictionary, a communication can even be simply an instance of *transmitting*, as in "the communication of disease". Therefore, it is important to be aware of what exactly the "communication" of plants involves, and not to incongruously model it in reference to our intentional forms of human communication. The same holds for other alleged plant capacities, such as associative learning, altruism, and memory.

Can plants act, and even act altruistically? Plants have amazing mechanisms to protect themselves from damage (at least to some extent). For example, when insects feed upon them, some plants emit chemicals that attract wasps, which in turn feed on the insects. Or the leaves that are fed upon release chemicals that signal to other leaves of the same plant to boost production of anti-feeding chemicals. Other plants in the neighborhood also receive these signals. This has been interpreted as plants warning each other or even as plants being altruistic. If altruism is to be understood as something like unselfish concern for the welfare of others, then this behavior is not altruistic because it is not unselfish, and it is not motivated by concern for the welfare of others. The behavior occurs unconsciously, similar to how our skin increases melanin production when exposed to sunlight. Therefore, while plants can do amazing things, they cannot *act* at all. Plants are not *agents*, in the sense that philosophers as well as behavioral and social scientists commonly use that term. Agency, in that context, usually refers to intentional action (Schlosser 2015). In contrast to automatic or reflexive behavior or activity, an *action*, as I will here use that term, is an intentional (and therefore purposive, conscious and subjectively meaningful) activity. Since plants are not conscious, they cannot have any intentions, purposes, desires or preferences, and nothing is subjectively meaningful to them. We can, of course, and we do *ascribe* purposes and preferences to plants, as we can and do ascribe them to bodily organs, bacteria or even toasters. But we need to be careful when using these ascriptions and remind ourselves that they are really just figurative talk – lest we fall prey to and encourage macroscopic ontological misunderstandings. There is overwhelming scientific consensus that plants cannot act and that they have neither feelings nor desires nor intentions.

Arguments based on specific accounts of welfare

An account of welfare tells us approximately what makes an individual's life go well or poorly. The focus is on *basic* (as opposed to merely derivative) prudential goods (that is, goods that are such because and insofar as they contribute to someone's welfare).[1] An example of a derivative good is money. Money is not good in itself, but it is good merely because of the various goods that it can bring us. Derivative goods derive their goodness from their relationship to basic goods. Accounts of welfare aim at identifying the basic prudential goods and, not to forget, the basic prudential bads. Monistic accounts of welfare posit only one good and one bad (for instance, pleasure and pain), while pluralistic accounts assume that there are several goods and/or several bads.

A welfare subject is an entity that can have positive or negative welfare. An entity's level of welfare depends on the presence of basic goods and bads in its life. To know whether plants are welfare subjects, we need to know what the basic prudential goods are. Only entities that are susceptible to these goods and bads (in this world) can be welfare subjects (in this world).[2] Classical utilitarians are hedonists about welfare: according to hedonism, pleasure is the sole basic good and suffering the sole basic bad.[3] Since plants are not able to suffer or to experience pleasure (in this world), they are – assuming the truth of hedonism – not welfare subjects (in this world). Preferentialism is another prominent account of welfare among utilitarians. According to this theory, an entity is doing well to the extent that its preferences are fulfilled, and it is doing poorly to the extent that its preferences are frustrated. Since plants cannot have any preferences (in this world), preferentialism also implies that they are not welfare subjects (in this world).

The concept of welfare can be distinguished from the concept of perfection. Sumner (1996) distinguishes perfectionist considerations from welfare considerations and argues that only the former apply to plants:

> Strictly speaking, however, what is at stake for non-sentient organisms, and what we are capable of affecting for better or worse, is not their well-being but their excellence or perfection as specimens of their kind.
>
> (Summer 1996, 179)

If Sumner is correct, plants are not subjects of welfare, but only subjects of excellence or perfection. Note that even knives are subjects of excellence or perfection, since they can be good specimens of their kind. In that sense, a knife can be a better or a worse knife and some things can be good or bad for a knife.

While Sumner's conceptual distinction between welfare and perfection is helpful, it is nevertheless possible to argue that plants are welfare subjects. Sumner's claim that they are not depends on his favored account of welfare, a version of hedonism. It is compatible with Sumner's conceptual distinction between welfare and perfection to argue that plants are welfare subjects (besides being subjects of perfection) and even to defend the substantive view that welfare consists in being a good specimen of one's kind.

Theories of welfare abound, and it remains controversial what the correct theory is (Bradley 2009; Fletcher 2015; Griffin 1986; Kraut 2009; Sumner 1996). In fact, it is controversial even whether one and the same theory applies to all welfare subjects, or whether instead different theories apply to different welfare subjects. May hedonism or preferentialism be true for humans, while a different theory of welfare is true for plants? Can we determine whether plants are welfare subjects (in this world) without committing to any particular account of welfare?

An argument that is not based on any particular theory of welfare

An argument against the view that plants are welfare subjects which does not depend on the truth of any particular theory of welfare can be summarized as follows:

P1: An entity is a welfare subject (in this world) if it can enjoy or suffer basic goods or bads (in this world).
P2: Normal human adults are welfare subjects (in this world).
P3: The same basic good(s) and bad(s) are good(s) and bad(s) for all welfare subjects.
P4: There are no basic good(s) and bad(s) that are good or bad for both plants and normal adult humans (in this world).
C: Plants are not welfare subjects (in this world).

Premise 1 of this argument is true by definition, and Premise 2 is uncontroversial. Premise 3 relies on the truth of welfare invariabilism, and Premise 4 relies on some considerations about health, life and flourishing. Let us now scrutinize Premise 3 and 4.

Welfare invariabilism, as I use the term here, refers to the view that the true account of welfare, whatever it is, is true for all welfare subjects. Welfare variabilism, in contrast, is the view that different basic good(s) or bad(s) are good or bad for different welfare subjects. Note that welfare invariabilism and variabilism are themselves not substantive accounts of welfare: they do not tell us what the goods and bads are.

Eden Lin (2018) defends welfare invariabilism. Lin presents two arguments in favor of invariabilism, the first of which appeals to the greater simplicity of this "one size fits all" approach to welfare: There is reason to favor a simpler theory over a more complicated one, other things being equal. Lin's second argument in favor of invariabilism is that it is hard to argue in favor of variabilism. This is because what, at first sight, may look like considerations in favor of variabilism are in fact not. I discuss these considerations below.

The first consideration that seems to favor variabilism appeals to the fact that some goods are inaccessible to some entities. Consider pleasure, which may well be a basic prudential good for people. Since pleasure is unavailable to plants, it seems that if something is to be basically good for plants, it cannot be pleasure. Therefore, it seems that we need different lists of basic prudential goods for different welfare subjects.

This can be denied, however. If pleasure is a basic prudential good, nothing precludes it from being a basic prudential good for plants. That pleasure is a good just means that if plants were experiencing pleasure, this would benefit them. This can be true even if, as it happens, plants do not experience pleasure. That pleasure is inaccessible for plants does not preclude it from being a good for them. Since any good could be a good for plants, there is no need to devise a separate list of goods for plants. More generally, the inaccessibility of some goods to some entities does not lend support to welfare variabilism (Lin, 2018).

The second consideration that seems to favor variabilism appeals to the different natures or capacities of different individuals. Certain goods, according to this line of argument, are only good for certain individuals with a particular nature. However, those who think that an individual's nature determines what is good for him or her typically accept a nature-fulfillment theory of welfare.

The ultimate good, according to this theory, is nature-fulfilment. On the nature-fulfillment theory, however, nature-fulfilment is the basic good for *all* subjects of welfare, whatever their nature happens to be (Lin, 2018).

One could object that on the view that welfare consists in nature-fulfilment, the lists of more specific goods for different individuals would differ, because different things would fulfill different individuals' natures. That is true. But it is also true that different things are pleasant for different individuals and different individuals desire different things. Nevertheless, hedonism accepts only one basic good, namely pleasure, and preferentialism accepts only one basic good, namely preference satisfaction. Similarly, nature-fulfilment is the sole basic good, according to nature-fulfillment accounts of welfare. It is irrelevant that nature-fulfilment consists in different things for different individuals, because the basic goods indicate what makes some state of affairs good: for example being an instance of pleasure, desire-satisfaction or nature-fulfilment.

Lin's defense of welfare invariabilism supports Premise 3 of the argument sketched above.

Given that normal adult humans are welfare subjects (Premise 2), and given that the same list of goods and bads applies to *all* welfare subjects (Premise 3), we need to know next whether anything at all is a basic good or bad for *both* normal adult humans and plants. It surely seems that some goods and bads are available to plants (in this world). Plants can live or die, they can be healthy or ill, and they can flourish or fail to do so. Normal adult humans, as well, can certainly be healthy and alive or unhealthy and dead (in this world), but none of these things actually is a basic prudential good or bad for normal adults (in this world) – or so I will argue. Furthermore, I will argue that flourishing is not a basic prudential good for normal adult humans (in this world).

Consider health first. Plants can certainly be healthy in a way. Gardeners aim to grow healthy plants, and they have criteria for determining whether any particular plant is healthy or sick. Indeed, the study of plant disease is a science – plant pathology – which explores the etiology of plant diseases and methods of monitoring or managing healthy plants. Disease, in this context, is defined as "sub-optimal plant growth brought about by a continuous irritant, such as a pathogen (an organism capable of causing disease) or by chronic exposure to less than ideal growing conditions" (Williams and Boehm 2017). For humans, being ill does not exactly mean the same thing: human health is not primarily a question of ideal growth. For example, regular headaches, nausea or psychiatric problems may not hinder my bodily growth, but they are still instantiations of diminished health. One might adopt a wider understanding of "growth" in the case of humans, but then it would no longer be the same thing as it is for plants. Furthermore, when it comes to an understanding of growth that encompasses non-bodily growth, it would be controversial and unclear what growth consists in and how it is supposed to relate to health. So, standard definitions of health for plants are not plausible definitions of health for humans. Likewise, some influential definitions of human health are quite broad and include items that make that concept of health inapplicable to plants. An example is the World

Health Organization's definition of health as "a state of complete physical, mental and social well-being and not merely the absence of disease or infirmity" (World Health Organization 2006). Be that as it may, the non-question-begging way of discussing this issue is by starting from a minimal definition of health, which is indeed applicable to both humans and plants: health as the absence of disease.

Being diseased is bad for plants, possibly a basic bad. But is being diseased also a basic bad for humans? One may argue that being diseased is bad, if it is, because of what it leads to. When we are diseased, we may feel bad in various ways. The illness may, furthermore, deprive us of certain pleasant or worthwhile experiences, and it may impede us from doing what we want or what would be good for us. If being diseased had none of these adverse effects, it would be hard to see why it would be bad for us. Being diseased, according to this argument, is not bad in itself. It is not a basic bad. It is only bad, if it is, because of what it leads to. It is merely derivatively bad.

The same, one may argue, holds for life. Life is good or bad, if it is, because of what it brings to us. A permanently comatose individual is alive. However, her life, under these conditions, does not seem to have any prudential value for *her*. It may be good or bad for *others* that this individual is still alive, but being alive is not in itself good or bad for her. This example suggests that whatever makes our lives good or bad for us are things that occur *within* our lives, and not being alive as such. Life, one may conclude, is a very important *derivative or instrumental* good, or bad, for humans: it is good or bad for what it brings or allows or leads to, and not in itself.

What about flourishing, then? Both plants and normal adult humans seem to be able to flourish, in a way. In addition, flourishing certainly sounds like something that is good in itself. However, what is flourishing? It better not turn out to be just a synonym for "welfare", as that would make it unhelpful for our purposes. After all, we want to know what constitutes welfare, not what other words we could use for saying "welfare".

Some authors refer to "flourishing" as a particular prudential good, perhaps the only good on the list of goods. This is the notion of "flourishing" that interests us in this paper. We need a better understanding of this alleged good before we can determine whether it is a final good for both plants and people. Richard Kraut is, perhaps, the most influential contemporary proponent of a substantive theory of welfare according to which welfare consists in flourishing. What is more, Kraut explicitly considers flourishing to be the prudential good for all welfare subjects. Welfare subjects, according to Kraut, are living organisms, such as plants, non-human animals and humans, and the final goods for them are plant, animal and human flourishing, respectively. Kraut (2009, 134) explains:

> If, as we have agreed, it is always good for a living thing to flourish, that is because the things in which flourishing consists are good for that living thing.

What, then, are the things in which flourishing consists? This is what we need to know in order to judge whether these things qualify as basic goods for both plants and humans. "For most living things", Kraut (2009, 5) notes, "to flourish is simply to be healthy . . .". This understanding of what flourishing consists in brings us back to the question whether health is a basic good. After all, if flourishing is good because of what it consists of, and it consists in health, then flourishing is a basic good just in case health is. Health for a human implies more than health for a plant, Kraut (2009, 90) argues: ". . . in our case flourishing consists not only in the healthy functioning of our bodies, but in psychological health as well". This statement is in agreement with more-than-minimal definitions of health, such as that of the WHO mentioned above; but what matters about the statement here is that it entails that, for both humans and plants, bodily health is a basic good. Against this claim, however, we encountered the argument that bodily health is merely *derivatively or instrumentally good* for humans. If so, our search for a *common basic good* for both plants and humans leaves us with empty hands.

One could challenge the argument exposed in this section in at least three ways. First, one could argue that bodily health and/or life is a basic good for normal adult humans. Second, one could present a plausible account of flourishing, perfection or nature-fulfilment that applies to both plants and humans. Third, one could argue that there is a list of basic goods that applies to both plants and humans. In order to make this last move more specific, consider a list of five goods and call the goods H1, H2, H3, P1 and P2. This list applies to both humans and plants. As it happens, the first three goods on this list are not available to plants in this world. One example of such a good may be pleasure. As it also happens, the last two goods on the list are not available to people in this world. Nevertheless, these goods are basic goods for people. This just means that if people were to realize these goods, this would be basically good for them. One example of such a good may be photosynthesis. If this were the case, both plants and people could be welfare subjects in this world, respecting welfare invariabilism. But are there any plausible examples of final goods for plants that are also final goods for people in this world? Photosynthesis itself does not look like a promising candidate, since it does not look like a basic prudential good for any entity. And I cannot think of any other potential good that would do the job.

Conclusions and implications

If, as most utilitarians believe, plants are not welfare subjects, then we have reason to treat plants in certain ways only in so far as our treatment of them affects the welfare of welfare subjects. The welfare of humans depends on plants greatly and in many ways: we need plants for food, clean air, water, fuel, clothing and shelter; many medications and things that we use in daily life are made from plants; many people highly appreciate plants in relation to recreational and spiritual activities, as well as cultural traditions; and the presence of plants has a significant positive impact on, and may even be indispensable for, our mental health and well-being. In addition to humans, also non-human animals – many of which are certainly welfare subjects – depend on plants in a number of these ways.

According to utilitarianism, we never have reason to care for plants at the (overall) expense of the welfare of humans or non-human animals. For example, we do not have reason to kill non-human animals in order to protect some plants, unless the protection of these plants has, in the long run, a greater positive effect on the welfare of human and/or non-human animals. According to utilitarianism, however, we often have reason to protect plants. After all, as mentioned above, plants happen to promote the welfare of welfare subjects in many ways.

Notes

1. The goods in question are also sometimes called "intrinsic", as opposed to "extrinsic" or "final", as opposed to "instrumental" goods (see Zimmerman 2015).
2. The clause "in this world" is necessary, because if the entities were susceptible to the goods and bads in some other possible world, they would be welfare subjects *in that other world*, but not in the actual world.
3. While I talk here as if the goods were *things*, such as pleasure, it is more accurate to talk of goods as being always *states of affairs*, such as the state of affairs that some individual, S, experiences pleasure.

References

Bradley, B. 2009. *Well-being and death*. New York: Oxford University Press.

Fletcher, G. 2016. *The Routledge handbook of philosophy of well-being*. New York: Routledge.

Griffin, J. 1986. *Well-being, its meaning, measurement, and moral importance*. New York: Oxford University Press.

Hogan, C. M. (Lead Author), Taub D. R. (Topic Editor). 2010. "Plant". In: *Encyclopedia of earth* edited by Cutler J. Cleveland. Washington, DC: Environmental Information Coalition, National Council for Science and the Environment. Last revised 15 August 2011. Accessed 27 September 2012, http://eol.org/info/449.

Kraut. 2009. What is good and why: The ethics of well-being. Cambridge, MA: Harvard University Press.

Lin, E. 2018. "Welfare Invariabilism." *Ethics* 128 (2): 320-345.

Liu, Q., and Lin, L. H. 2017. Organ-to-organ communication: a drosophila gastrointestinal tract perspective. *Frontiers in Cell and Developmental Biology*, April, no. 5: 29.

Schlosser, M. 2015. "Agency", *The Stanford Encyclopedia of Philosophy* (Fall 2015 edition), Zalta, E. N. (ed.). Accessed 28 October 2017 from https://plato.stanford.edu/archives/fall2015/entries/agency/

Sumner, L. W. 1996. *Welfare, happiness and ethics*. New York: Oxford University Press.

Williams, S. and Boehm, M. J. 2017. Plants Get Sick Too! An Introduction to Plant Diseases. Fact Sheet. Ohio State University College of Food, Agricultural, and Environmental Sciences. Accessed 28 December 2017 from https://ohioline.osu.edu/factsheet/plpath-gen-1.

Wohlleben, P. 2015. *Das geheime Leben der Bäume. Was sie fühlen, wie sie kommunizieren – die Entdeckung einer verborgenen Welt*, Ludwig Verlag, München.

World Health Organization. 2006. *Constitution of the World Health Organization – Basic Documents* (Forty-fifth edition), Supplement, October 2006. Accessed 28 December 2017 from www.who.int/governance/eb/who_constitution_en.pdf.

Zimmerman, M. J., 2015. "Intrinsic vs. extrinsic value", *The Stanford Encyclopedia of Philosophy* (Spring 2015 edition), edited by N. E. Zalta. Accessed 28 December 2017 from https://plato.stanford.edu/archives/spr2015/entries/value-intrinsic-extrinsic/.

3 Is considering the interests of plants absurd?

Ronald Sandler

Introduction

It is clear that individual plants often have *instrumental value*. Plants provide goods and services or are useful as a means to an end. They offer food, shade, materials, recreation, aesthetic experiences, and medicine. Some people hold the view that plants also have *inherent worth*. They believe that plants have interests, that moral agents ought to care about them for their own sake, and that wantonly harming or killing them is therefore wrong. In environmental ethics, the view that all plants (indeed, all living things) have inherent worth is often called *biocentrism*. (Biocentrism is the view that each plant has its own worth or value; whether species or populations of plants have inherent worth is another matter.)

Other people believe that individual plants have only instrumental value. Many ethicists who hold this view believe that ascribing inherent worth to plants has unacceptable implications. They believe the ascription carries the implausible implication that other entities, such as microbes and garden hoses, have inherent worth and/or that we cannot use plants for our own ends, which we clearly must do in order to survive. In other words, these ethicists offer *reductio ad absurdum* arguments against the view that plants have inherent worth.

In this chapter, I defend the view that plants have inherent worth against the criticism that it has absurd implications. In the first half of the chapter, I present the case in favour of recognizing that plants have inherent worth. In the second half of the chapter, I respond to *reductio ad absurdum* objections to the view that plants have inherent worth.[1]

The case for the inherent worth of plants

Inherent worth, as the term is being used in this paper, is a type of value or way of mattering. To say that something has inherent worth is to claim that (1) it has interests or a good of its own, and (2) moral agents should care about its interests for its own sake (Taylor 1986). If something has inherent worth, then its interests need to be considered in deliberations regarding actions, practices and policies that affect it.

That plants have interests or a good of their own seems to many people to be clearly true. After all, it is possible to harm plants by, for example depriving them of nutrients or destroying their roots. People commonly speak about what is good or bad for the plants in their gardens; farmers work to create conditions that are favorable for their crops; and conservation biologists try to help threatened plant specimens thrive. If it is possible to impair or benefit something's health or wellbeing, then there must be things that are and are not in its interests, according to this view (Attfield 1991).

One possible response to this commonsense view is that plants do not have interests because they cannot experience anything: "The capacity for suffering and enjoying things is a prerequisite for having interests at all, a condition that must be satisfied before we can speak of interests in a meaningful way" (Singer 1989, 154). However, why should this be? After all, it is a common view that a person's welfare depends on more than just their psychological states and experiences. It also depends upon such things as their bodily health. If this is right – or even possibly right – then there seems no conceptual problem with the idea that plants could have a biologically, but not psychologically, based welfare.

A reason often given for the view that nothing could make a difference to a nonconscious thing's welfare (or that it does not have a welfare at all) is that nothing could matter to it. Nonconscious things, because they are nonconscious, cannot care about what happens to them. That is, they cannot take an interest in anything, including what is done to them, precisely because they lack the capacity for attitudes, intentions and desires. However, this reasoning seems to depend on an equivocation. "X has an interest in Y" can mean either "X is interested in Y" or "Y is in X's interest." Therefore, to say "oak trees have an interest in not contracting oak wilt fungal disease" can be understood either as "oak trees are interested in not contracting oak wilt" or "oak wilt is not in an oak trees' interests." The first of these statements is absurd. Oak trees do not "take an interest" in anything, since they lack mental states. However, this is not the same as claiming that oak wilt is not in their interest. Once the ambiguity of "having an interest" is removed, it seems that the fact that trees cannot take an interest in anything does not preclude their having interests.

Proponents of the view that plants lack a welfare might respond by asking: If *they* (the plants) cannot take an interest in their welfare, why should their welfare matter to us? However, this response actually concedes that plants have interests, and changes the issue to whether we ought to care about them (a topic discussed below). Moreover, there actually is a sense in which plants take an interest in their welfare, albeit a nonconscious one. They do things that involve actively pursuing what is beneficial to them. For example, they defend themselves against predators, repair damage, fight disease and have phototropic leaves and skototropic roots.

The discussion above concerns biocentrism's response to the view that consciousness is necessary for having interests. However, for biocentrism to be justified, it is not sufficient to show that consciousness is not necessary for interests. It must also be demonstrated that nonconscious living things possess interests. If their interests are not based on psychological experiences and attitudes, where do they come from? As mentioned earlier, the concept of interests is connected with that of benefit and harm. The general idea is that a thing has interests if we can talk meaningfully about benefiting and harming it directly, or without reference to anything else's interests. Therefore, in order to establish positively that nonconscious living things have a good of their own, it is necessary to provide an account of what grounds their interests. That is, there must be an explanation for why Asian long-horned beetles are bad for trees, whereas sunlight is good for them.

The most common explanation for what grounds the interests of nonconscious organisms (as well as the bare biological interests of conscious ones) is that the organisms are *goal-directed systems* (Goodpaster 1978; Sandler 2007; Taylor 1986; Varner 1998). Their parts and processes are organized in ways and for reasons related to accomplishing things such as resisting predators and reproducing. There are "in order that" explanations for why it is that they have the parts that they do, the processes that they do, and the life cycles that they do. Flowers produce nectar *in order to* attract pollinators; ivy has tendrils *in order that* it can climb; tree roots are skototropic in order that they grow deeper into the ground.

That nonconscious organisms are end or goal-directed systems means that it is possible to affect them in ways that are conducive or detrimental to accomplishing their goals or ends – that is to harm and benefit them. Asian long-horned beetle larvae, which bore into the bark of elm trees, are bad for the trees because they diminish the bark's capacity to distribute water and nutrients throughout the tree's living tissue, a process that is needed for the tree to accomplish its ends of survival and reproduction. This is a *teleological account of interests*.

The explanation for why the living tissue of elm trees has the function of nutrient distribution is that the tissue was selected for because it performed this role in prior generations of elm trees. The trait has persisted – it exists in current elm trees – because it performs that function in the elm tree form of life. In the case of naturally evolved organisms, the selection etiology or causal process by which the teleology of their parts and processes is determined is natural selection. This is an *etiological account of teleology*.

The teleological account of interests and etiological account of teleology combine to produce the following argument for the view that plants have interests:

1. An entity has interests if it is possible to benefit and harm it directly. (Definition of interests)
2. It is possible to benefit and harm an entity directly if it is goal-directed, that is, if there is something it is oriented toward accomplishing. (Teleological account of interests)
3. In plants, goal-directedness is accounted for by the explanations for why their parts and processes persist in organisms of that type. (Etiological account of teleology)

4. Therefore, the interests of plants can be specified in terms of what resources, conditions, and treatments are conducive to, or detrimental to, the realization of their goals (e.g. survival and reproduction) in the ways typical of organisms of that type.

This argument establishes only that plants have *biological interests*. The argument does not imply that they have the same interests as sentient nonhumans or as people, or that their biological interests should be taken into account to the same extent or in the same way as the *psychological interests* of sentient nonhumans and people. In fact, the argument does not even establish that the biological interests of plants should be taken into account at all. That is, it does not show that they are morally considerable or have inherent worth.

As mentioned earlier, it is uncontroversial that we should care for plants for instrumental and other human-oriented reasons. For example, trees often provide ecosystem services, are used as natural resources, are crucial to maintaining well-functioning ecological systems, are people's property, have economic value and have cultural importance. However, should we care about their interests *for their own sake*, such that we have duties and responsibilities directly to them and not only regarding them?

Proponents of biocentrism commonly use an extensionist approach to argue that we ought to care about the good of all living things. This approach involves showing that there is no adequate justification for caring about the interests of sentient (or conscious) organisms but not the interests of nonconscious organisms. There are several interrelated strategies for trying to motivate this. One is to apply a principle that Singer (1975, 1989) uses when arguing for the inherent worth of sentient animals: *if something has interests, then the presumption is to consider them, unless a good justification is provided for why they ought not be considered.* On this view, the absence of a good reason not to consider the interests of an entity, including plants, is sufficient to establish that they are morally considerable and have inherent worth.

A second extensionist strategy is to argue that if we ought to care about the biological interests of people and sentient nonhuman animals, independent of how it affects their psychological experiences, then we ought to care about the biological interests of nonconscious organisms (Varner 1998). Imagine that a person is being slowly poisoned. However, the poisoning is not causing them any psychological distress or pain, since the damage is occurring at a very low rate, and they have no knowledge of it. If the poisoning is nevertheless wrong, this would seem to imply that bare biological interests matter, independent of the psychological impacts.

Another approach to justifying direct consideration of plants is coherentist. For example, Taylor (1986) argues that consideration of the interests of all living things best coheres with a proper understanding of humanity's relationship with the natural world. Taylor defends a set of empirical claims or beliefs that describe the many ways in which human beings are similar to organisms of other species. We, like them, are evolved biological organisms that have physical requirements for survival and wellbeing. We, like them, depend for our wellbeing on the natural world, which is itself a system of interdependence of which humans and other species are a part. We, like them, are living things with a good of our own. Taylor believes that once a proper biological, ecological and evolutionary understanding of *Homo sapiens* and other species is accomplished, we will see that the most reasonable view – the one that best coheres with that well-informed perspective – is that all living things have inherent worth regardless of their species membership.

Taylor also argues that several historically common claims regarding unique human abilities have turned out to be mistaken. Human beings, for example, are not the only ones who communicate through language, use tools, teach their offspring, solve novel problems, are social, enforce group "rules" or are altruistic. In an evolved biological world such as ours, differences between species tend to be matters of degree. Of course, differences in degree can be significant. So far as we know, no other species innovates, disseminates and accumulates knowledge, ideas, technologies or social structures at the rate or on the scale that humans do. There is thus quite a lot that is distinctive about human beings, even if we are in an ecological and evolutionary sense plain members of the biotic community. However, being distinctive is nothing unique to us. Every species is distinctive; otherwise, it would not be a separate species. We are excellent at mathematics, developing technology, innovating languages, and building complex social systems, in comparison with individuals of other species, and our capacities are particularly important to our cultural form of life. However, other species' capacities are just as important to their forms of life. Buoyancy control is crucial for salmon; silk spinning is crucial for golden orb spiders; and photosynthesis is crucial for silver maples. This leads Taylor to ask: Why should the traits that are crucial to the form of life of one species be the standard by which we judge the worth of the others? He argues that it cannot be that cultural capacities or robust agency are the standard because they are crucial to us, since this assumes that we have special status in order to prove that we do. He therefore concludes that there is no non-question begging justification for asserting the worth of individuals of one species while denying the worth of individuals of other species, *Homo sapiens* included (Taylor 1986).

The extentionist and coherentist approaches to justifying the direct moral considerability of plants are generally regarded as complementary. If they are successful, they establish not only that plants have interests but also that their interests need to be considered for their own sake.

Objections and replies

A number of objections have been raised to the idea that plants are directly morally considerable. Many of them involve claiming that the view itself is absurd or else has absurd implications. In this section, I review the most prominent of these objections and outline how proponents of the inherent worth of plants can respond to them.

Objection one

The view that plants have equal (or the same) inherent worth as people is absurd, since it is clearly permissible to do all sorts of things to plants, such as own them, that is not permissible to do to people (Schmidtz 1998).

Response to objection one

This objection assumes that the view that plants have equal (or the same) worth as people implies that plants should be treated in the same way as people. However, this inference is not valid. One important reason for this is that the claim that plants are directly morally considerable underdetermines how they ought to be considered. For example, one way to understand equal worth, often associated with utilitarianism, is for like interests to be considered alike (Singer 1975, 1989). However, plants and people have very different interests. People (and some nonhuman animals) have psychological interests in addition to biological ones. On many deontological accounts, including Taylor's (1986), the fact that people have complex cognitive and psychological capacities that plants lack entitles them (and not plants) to rights. On many virtue ethical accounts, plants might be due consideration, but they will not be due the compassion appropriate for sentient beings or the respect due to robustly rational beings because plants lack the relevant capacities (to suffer and to be moral agents, respectively) for those forms of responsiveness to be appropriate (Sandler 2007). It is beyond the scope of this chapter to discuss in detail the variety of ways in which normative theories translate inherent worth into particular forms of moral standing and responsiveness. Suffice it to say that consequentialism, virtue ethics and deontological ethics each provide resources for responding to this objection (Sandler 2018).

Another reason that this objection is unsuccessful is that it is possible that plants have inherent worth but not equal inherent worth (or the same inherent worth) to that of people. Perhaps the fact that plants are goal-directed systems justifies their being regarded as having some worth but not as much as entities with a broader range of capacities or more complex capacities.

A third reason why this objection is unsuccessful is that it conflates equal consideration with equal treatment (Singer 1975, 1989). Even if people and plants have equal worth, it does not follow that they should be treated in the same ways. How to promote an organism's welfare depends upon the facts about its form of life. Pigs and dolphins might both have inherent worth and have it equally, but it would be wrong to release dolphins into the field and pigs into the ocean. Equal worth and equal consideration do not imply same treatment. That is to say, two entities might have the same type of value (inherent worth), such that their interests need to be considered in the same way or to the same extent (equal consideration), and yet they should be treated quite differently. An adult human and an infant human both have inherent worth, and their interests should therefore be considered for their own sake, but we should not treat them the same.

Objection two

The view that plants have inherent worth is absurd since we need to eat them to survive and to kill them to protect ourselves (Williams 1995).

Response to objection two

This objection commits several of the same mistakes as the first objection. As discussed above, ascriptions of inherent worth to plants underdetermine how to take them into consideration and what the practical implications are. It is consistent with plant interests being morally considerable that they are less normatively significant or differently significant than those of rational beings or sentient beings. It is also consistent with there being other types of values that need to be taken into consideration and that can take precedence over plant interests. Again, this is true on a wide variety of normative theories, including consequentialist, deontological and virtue-oriented theories (Attfield 1999; Carter 2005; Sandler 2007; Taylor 1986). Therefore, it does not follow from plants having inherent worth that it is always wrong to kill them or use them for our purposes, particularly when our serious or basic interests are at stake. In fact, on a minimalist conception of the normative significance of plant interests, all that consideration of them would require would be that plants not be destroyed or harmed wantonly (or for no reason at all), or perhaps that plants not be destroyed or harmed without acknowledgment or the recognition of a loss.

Objection three

The view that plants have inherent worth is absurd, since it would imply that every single organism has inherent worth as well, including bacteria, and we cannot avoid relentlessly killing microbes with our immune systems (Davion 2006).

Response to objection three

The etiological account of teleology and teleological account interests do imply that bacteria have a good of their own. However, it does not immediately follow from this that human beings ought to care about their good or take their good into consideration. Having a good of one's own may be necessary but not sufficient for inherent worth (Feinberg 1974; Williams 1995). If there are good, non-arbitrary reasons not to extend direct moral consideration from plants to microbes, then this objection can be avoided. There are such reasons. Even if microbes have interests, for practical purposes we are not moral agents with respect to them, due to their scale and multitude. It is not possible for us to consider their individual interests; therefore, we do not need to. Moreover, as indicated above, inherent worth underdetermines how something is to be taken into account. Therefore, even if we should regard them as having inherent worth, and so should not wantonly destroy them, this might have no practical implications. We are occasionally faced with the question of how to treat a particular tree or plant, but very rarely (and under extremely specific circumstances) are we faced with the question of how to treat a particular microbe.

Objection four

The view that plants have inherent worth is absurd since it would imply that all kinds of artifacts, such as wheelbarrows and thermostats, have inherent worth, as well (Kaufman 1994; Thompson 1990).

Response to objection four

According to the teleological account of interests, an entity can be benefited or harmed directly if it is a goal-directed system. According to the etiological account of teleology, an entity is goal-directed if the causal explanation for why it has the parts and processes that it does is that they were selected because of the role they perform in bringing about certain ends. These two taken together establish that plants have interests. However, they seem to imply that anything that is a goal-directed system as a result of a selection etiology would likewise have interests. Intentionally created functional artifacts are teleologically organized, and there are etiological explanations for their parts and processes. There is a reason that wheelbarrows have two handles set apart the width of a human body, a single front wheel and two rear feet. This design makes it easier to load and move materials. The parts and processes were selected for (or designed) and persist in current wheelbarrows because they are conducive to accomplishing that end (moving materials). The sensor in a thermostat is there in order to regulate temperature. The axles of a car are there in order to generate locomotion. The fan inside a computer is there in order to keep it from overheating. The sharp edge of a knife blade is there in order to cut effectively. Therefore, this objection is correct that an implication of the teleological account of interests and etiological account of teleology is that most artifacts also have a good of their own. (Here is another way to think about this: the work of grounding the interests of plants is not done by the "natural" (or biological) component of "natural selection", but by the "selection" component.) This is a surprising and counter-intuitive implication. However, it is not in itself absurd. As with microbes, it does not follow from this alone that artifacts have inherent worth or that we cannot harm them. Again, the reason is that something's having interests underdetermines whether we should take its interests into account, how we should do so, and what the implications of doing so are. For example, several of the considerations that Taylor (1986) argues justify regarding plants as directly moral considerability, such as common origins with other species and ecological interconnectedness, do not apply to artifacts. Moreover, even if (nonconscious) artifacts have inherent worth, perhaps our interests always take precedence over theirs, given their relative simplicity and that their interests are grounded in our designs and intentions with respect to them. Thus, while it might be strange to think about (nonconscious) artifacts as having a good of their own, it does not necessarily lead to absurd conclusions or implications.

Objection five

The idea that plants have inherent worth is absurd since there are so many of them that it is not possible to take all of their individual interests into account (Thompson 1990).

Response to objection five

There are many plants. However, there are also many people. A multitude of a kind of thing is not itself sufficient for justifying the claim that it is not possible to take the interests of individuals of that kind into account. As observed earlier (Objection Three), we are often faced with questions about what to do with a particular tree or a small number of plants. Plants are more similar to people than microbes in this regard. In cases where very large numbers of plant interests are involved and it is not possible to consider them individually, which occurs frequently with respect to people and nonhuman animals as well, proponents of the inherent worth of plants often advocate acting in ways that are conducive to the health of the system (Taylor 1986). The idea is that doing so will be the best way to respect or promote the interests of the most individuals. This strategy is an instance of the use of secondary principles in ethical decision-making (Mill 2001). Thus, the challenges here do not appear to be any different in kind than those faced by normative theories on sentientist or anthropocentric accounts of inherent worth, and the same resources to address them can be employed.

Objection six

The view that we should take plant interests into consideration is absurd because it is not possible to know what their interests are, since they cannot communicate them to us or represent them in decision-making processes (Baxter 1975).

Response to objection six

It is true that plants cannot speak, but it is not true that we cannot know their interests. Through experience and observation, gardeners, naturalists, farmers, conservation biologist and others are able to identify the conditions that are conducive or detrimental to individuals of a plant species thriving. Acid rain is clearly bad for trees because it damages their foliage, which is crucial to their capacity to grow and reproduce. Standing water and damp ground is clearly good for phragmites, since those are the conditions under which they can survive, grow and spread. Moreover, the fact that plants cannot engage in collaborative decision making with us is not a barrier to considering their interests. Pets do not participate in discussions regarding treatment options in veterinarian offices, but their interests are nonetheless considered. Endangered species do not participate in discussions regarding recovery plans for them, but their interests are considered. Determining the interests of plants is possible, and taking their interests into consideration is in fact common; therefore, this objection fails.

Conclusions

The view that plants have inherent worth is not absurd and does not have absurd implications. Those who believe that it does generally fail to distinguish between (1) accounts of which entities have interests, (2) accounts of which entities have inherent worth, and (3) accounts of the practical implications of an entity's having inherent worth (which is a matter of normative theory as well). Once these are distinguished, it is clear that plants having inherent worth does not imply, for example, that they have rights, that they cannot be used for human purposes, that they should be treated (or considered) in the same way as people, or that they are too numerous to have their interests considered.

The most common and promising approach to establishing that plants have inherent worth is one that relies on an etiological account of teleology and a teleological account of interests. The combination of these accounts does have some surprising implications, such as artifacts having a good of their own. However, this is not in itself absurd, nor does it lead to absurd implications, for the reasons discussed above. Moreover, the alternative is to deny either the etiological account of teleology or the teleological account of interests (or accept a strictly mentalistic account of welfare), which leads to an even more surprising conclusion —that is, plants do not even have interests.

If the arguments in this chapter are correct, then there are promising arguments in favor of plants having inherent worth that do not expose the view to *reductio ad absurdum*. It is possible for plants to have inherent worth and for us to eat them too.

Acknowledgments

The author would like to thank John Basl and Marcello Di Paola for their helpful comments and suggestions on an earlier draft of this manuscript.

Note

1. Some of the material in this chapter is adapted from Sandler (2018) and Basl and Sandler (2013).

References

Attfield, R. 1991. *The ethics of environmental concern*. Athens, GA: University of Georgia Press.

Attfield, R. 1999. *The ethics of the global environment*. Edinburgh, UK: Edinburgh University Press.

Basl, J., and R. Sandler. 2013. The good of non-sentient entities: Organisms, artifacts, and synthetic biology. *Studies in History and Philosophy of Biological and Biomedical Sciences* 44, no. 4: 697–705.

Baxter, W. F. 1975. *People or penguins: The case for optimal pollution*. New York: Columbia University Press.

Carter, A. 2005. Inegalitarian biocentric consequentialism, the minimax implication and multidimensional value theory: A brief proposal for a new direction in environmental ethics. *Utilitas* 17, no. 1: 62–84.

Davion, V. 2006. Itch scratching, patio building, and pesky flies. *Environmental Ethics* 28, no. 2: 115–28.

Feinberg, J. 1974. The rights of animals and unborn generations. Pp. 43–68 in *Philosophy and environmental crisis*, edited by W. Blackstone. Athens, GA: University of Georgia Press, Athens.

Goodpaster, K. E. 1978. On being morally considerable. *The Journal of Philosophy* 75, no. 6: 308–25.

Kaufman, F. 1994. Machines, sentience, and the scope of morality. *Environmental Ethics* 16, no. 1: 57–70.

Mill, J. S., and G. Sher. 2001. *Utilitarianism*. Indianapolis, IN: Hackett Pub.

Sandler, R. 2018. *Environmental ethics: Theory in practice*. New York: Oxford University Press.

Sandler, R. L. 2007. *Character and environment: A virtue-oriented approach to environmental ethics*. New York, NY: Columbia University Press.

Schmidtz, D. 1998. Are all species equal? *Journal of Applied Philosophy* 15, no. 1: 57–67.

Singer, P. 1975. *Animal liberation: A new ethics for our treatment of animals*. New York: New York Review.

Singer, P. 1989. All animals are equal. Pp. 148–62 in *Animal rights and human obligations*, edited by T. Regan and P. Singer. Upper Saddle River, NJ: Prentice Hall.

Taylor, P. W. 1986. *Respect for nature: A theory of environmental ethics*. Princeton, NJ: Princeton University Press.

Thompson, J. 1990. A refutation of environmental ethics. *Environmental Ethics* 12, no. 2: 147–60.

Varner, G. E. 1998. *In nature's interests? Interests, animal rights, and environmental ethics*. New York: Oxford University Press.

Williams, B. 1995. Must a concern for the environment be centred on human beings? Pp. 233–40 in *Must a concern for the environment be centered on human beings?* edited by B. Williams. Cambridge, UK: Cambridge University Press.

4 The flourishing of plants

A neo-Aristotelian approach to plant ethics

Angela Kallhoff

Plants have a range of specific characteristics. They grow throughout their life. They have a big surface and they are territorial creatures. They only need water, chemicals, and sunlight to survive (Larcher 2001). Plants are also useful resources for humans. They are the basis of the food chain. In addition, plants are also very beautiful. Plant life delivers a rich diversity of forms and of colors. Processes of plant self-development and the ever changing forms are elements of an extraordinary aesthetic experience (Böhme 1989).

In this chapter, I wish to combine these insights into the value of plant life with another perspective on plant life. Under non-detrimental circumstances, each single plant has the capacity to flourish. The concept of "flourishing" has gained momentum in the re-interpretation of Aristotle's notion of the good life of beings (Cooper 1986, 89). Following Aristotle, the concept of flourishing is the measure of the good life of all beings that have a soul, where a "soul" is the principle of life. Living beings as diverse as humans, animals, and also plants have souls, and each of these beings possesses a distinct range of innate capacities. All living beings that have souls are also capable of flourishing: of developing their own good life according to a set of innate capacities.

Sorabji explains this Aristotelian extension of the concept of a soul to plants:

> But how, then, does he [Aristotle] justify continuing to attribute a soul to plants? By extending the concept of soul, so that the non-conscious processes of nutrition and growth will now count as an activity of the soul. This extension may sound strange to us. But appeal to a (non-conscious) soul is needed, Aristotle thinks, to do justice to such facts as that a plant does not expand haphazardly, but preserves, or develops, a certain distinctive organization.
>
> (Sorabji 1975, 44)

New findings in contemporary botany contribute to the insight that plants have innate capacities and develop what might be called a "distinctive organization". In particular, in taking into account these findings, a two-sided account of the "good life" of plants can be established (Kallhoff 2002; 2014a). On the one hand, flourishing has an empirical side: a plant that flourishes fulfills its particular life-cycle; it is vital and copes successfully with episodes of external stress throughout the life-cycle; and it realizes a set of species-specific capacities also visibly expressed in a specific morphology. On the other hand, flourishing has an evaluative side: it is a concept that indicates the "good life" of a plant. Although this "good" is a pre-moral notion (Frankena 1979, 1–10), it is in important respects comparable to notions of welfare in other living beings. At a minimum, effects on plants that support flourishing or are detrimental to flourishing can be categorized as "beneficial" versus "harmful", whereas "beneficial" is preferable, when not outweighed by other, more important considerations.

A good life is the best situation that a single living being can realize. As for plants, this does not amount to a life without episodes of stress or duress, as plants need stress in order to develop strength (Larcher 2001). A plant's flourishing makes for a good life of a plant, in that the plant succeeds to accomplish its life-cycle and thereby realizes a full range of capacities, including proliferation, without suffering life-impeding harm. In short, flourishing is an empirical-cum-evaluative concept of plant life.

The concept of flourishing is helpful in plant ethics for three reasons. First, this concept has a distinct empirical content, so that it provides a yardstick for assessing which causal effects are harmful to a plant and which ones are beneficial. Authors who argue against plant ethics usually also presuppose that plants cannot experience harm. This reservation can now be rejected: even though plants do not feel pain, flourishing provides a theoretical frame for assessing harm to plants.

Second, even though flourishing is a pre-moral notion of the good life of a being, such notion carries evaluative weight. Plant ethics argues that humans have an interest in the flourishing of plants. Yet, besides this, it also matters to the single plant whether or not it succeeds in flourishing. In order to avoid anthropomorphism from the beginning, I do not wish to frame this in terms of "concern" or "interest" of the plant. In my best understanding of plant life, plants do not feel pain, nor are they conscious in any way comparable to animals and humans (Larcher 2001). But since plants invest in their flourishing, it apparently matters to them whether or not they flourish.

Third, neo-Aristotelianism helps to discuss plant ethics anew in one further respect. Neo-Aristotelians are not only interested in a concept of the good life that is closer to the characteristics of the living organism than are preference-based and desire-based accounts of the good life. Neo-Aristotelians also discuss concepts of a good life as part of at least a temporary vision of peaceful co-existence of a variety of different types of living beings. A neo-Aristotelian plant ethics thus provides not only new ways to generate a more comprehensive vision of the good life than one reserved to humans and non-human animals alone. It also underscores the possibility of peaceful and fruitful co-existence among different beings, as it sees the flourishing of one entity as not necessarily detrimental to that of other living beings that flourish according to their capacities. Instead, a "temporary cosmos" is envisioned: a scenario in which all living beings succeed in mutually supportive, at least non-destructive co-evolution.

This chapter unfolds as follows: the first section outlines the concept of flourishing in its application to plant life. The second section argues that the flourishing of plants matters in ethics, and provides a short sketch of the background of what I labeled "plant ethics" in an investigation of the concept of "flourishing" and principles of moral concern that relate to that insight (Kallhoff 2002). The third section proposes to distinguish various types of human–nature relationships. In each of these spheres, flourishing has a different moral meaning. The conclusion highlights further perspectives in plant ethics that work with a neo-Aristotelian approach.

Plant flourishing

Plants differ radically from humans and non-human animals. In order to give an empirically correct assessment of plant life a first, obligatory step is to consult botany (Larcher 2001). Plants react in a multitude of different ways to their environments. As sedentary organisms and due to a particularly big surface, they are confronted with what botany classifies as "stressors" of all kinds (Hirt 2009). Studying the reaction to stress in plant life helps us to understand what plants do and how they live. In particular, capacities to discern effects that result from different surroundings, and capacities to eject chemicals that serve as signals to other plants, have been studied (Heil and Karban 2010).

Three central aspects of plant life are remarkable. *First*, plants are units that respond systematically to stimuli. In this context, botanists also employ the concept of a "strategy": in order to survive in an ever changing environment, plants employ strategies that guarantee the continuation of this functional unit. *Second*, plants organize themselves in ways that support not only their long-term survival, but also the completion of a life-cycle and the continued exercise of a multitude of metabolic activities, leading to proliferation at least once in a lifetime. *Third*, plants strive to realize morphological and functional structures that are characteristic of the species and of members of the kingdom of plants. In short, a plant flourishes when it realizes a full life-cycle, when it develops its characteristics during this process, and when its vitality (its continued stress-responsiveness without loss of capacities to react strategically and without illness) is kept intact.

In this interpretation, a plant is still a unit that develops through time by transforming energy and chemicals into new materials. But instead of interpreting plants exclusively as functional units, as "survival-machines", an Aristotelian interpretation states that each plant has a set of plant-capacities, including nutrition, growth, and proliferation. These capacities are innate and a plant thrives when realizing them in specific ways, which include a distinct morphology.[1] Throughout the life-cycle, a plant struggles with aversive external stressors in order to fully realize the shape and the basic functional units that can express its innate capacities in various phases of its life-cycle. The evaluative side is provided by assessing these processes with reference to a description of what deserves respect in a moral framework. An argument for the moral significance of flourishing in a moral framework is given in the next section.

The assessment of damage and harm to plant life

The first important advantage of placing emphasis on flourishing is that this notion offers ways to define "harm" to single plants. When it comes to plants, harm indicates causal effects providing obstacles to flourishing. By poisoning a plant or sabotaging its habitat, we make it difficult (or at least more difficult) for a plant to flourish. Yet, we cannot conclude from this that harming a plant is morally wrong: we must also address the question of why harm to a plant or the opposite matter and if so in which way they matter. A first insight is provided by discussing the meaning of "good" in the context of harmful or beneficial causal effects on living beings.

Georg Henrik von Wright (1972, 9) explains that there is a special meaning of "harmful" when applied to the life of living beings: it is the opposite of effects on living beings that are termed "good for" that being. Calling something "good for" or "beneficial to" the life of a being does not only presuppose that the being is capable of doing better or worse, but also that it is doing better or worse has an axiological component. By calling something "good for" that being we do not only mean that there is a certain effect on that being, but we also evaluate that effect as being positively correlated with the good of the being. Von Wright states that there necessarily is an axiological component involved when addressing a beneficial causal effect on a living being:

> The other component . . . concerns the relations of these effects [useful or beneficial effects] to that which we have termed the good of a being. This is not a causal relation. It is more like a relationship of belonging.
>
> (Von Wright 1972, 49)

Note that this is a different way to explore the meaning of "good" than proposed by biocentric authors who relate the good of the being directly to the capacity of that entity to live (see e.g. Attfield 1981; Taylor 1986). In particular, I do not employ the concept of a "value", as the application of that concept to natural beings is difficult and controversial (see Pellegrino in this volume). Instead, the common usage of "beneficial" and "harmful" reveal that in generating effects on plant life, is it not only their well-being that is at stake. Instead, the assessment of these effects already implies that beneficial effects are "good for" that being in terms of enhancing something that should be taken into account, namely its flourishing; and harmful effects are "bad for" that being in providing obstacles to flourishing. This also proposes that beneficial actions are prima facie unproblematic, perhaps even desirable, whereas harmful actions need to be assessed in a framework in which "harm" needs to be debated anew.

Today, we know that plants are much more complex and responsive beings than generally presupposed. It has been argued that plants are particularly intelligent beings that build intelligent networks with their roots (Mancuso and Viola 2015). It has also been argued that plants can communicate via airborne signals (Heil and Karban 2010) and via intelligent root systems (Johnson and Gilbert 2015). Some have even argued that plants have mnemonic and learning capacities (Affifi 2013). Even granting all that, though, it remains true that plants differ significantly from humans and animals. Yet, the argument from a lack of complexity is fruitless. Moreover, respect regarding the flourishing of individual plants means attention to the distinction between harmful and beneficial effects on the lives of plants. Different from a value-model and different from a status-model in ethics, the main reason for claiming moral attention to flourishing is not that plants are welfare subjects, even though this assessment can be right when "welfare" is explicated in terms that fit the life of plants. Instead, what matters is whether or not actions and their consequences turn out to be harmful or beneficial.

My proposal to profit from a neo-Aristotelian interpretation of plants gives space for various types of harm to plants. In particular, supporting and respecting the flourishing of plants is in many ways not only desirable but also in the deep interest of humans. It is flourishing apple trees that provide the apples! Since the flourishing of plants is coupled with a full life-cycle of plants and with proliferation, humans who live from fruits, vegetables, and parts of plants need plants that flourish.

Philosophers usually also ask another question: They want to know whether or not causal effects matter to the plant itself. This question does not have to be decided in a neo-Aristotelian framework. There is, however, some evidence that even this might be the case: *First*, even though we do not know what it is like to be a plant,[2] it might matter to a single plant whether or not it flourishes. Plants show signs of suffering from stress; when confronted with overwhelming stress and harm they die. Even though this does not hurt them, it is bad to the plant. *Second*, the crucial question in plant ethics is whether or not we or certain other actors should care about the effects of actions on the flourishing of plants. This is the choice of constructive behavior regarding plants as opposed to destructive behavior. Even though it is not the space here to outline the underlying moral framework in detail (see Kallhoff 2002), it can be argued that constructive behavior is choice-worthy and morally recommended. Introducing this insight in the assessment of effects of actions and processes on nature is an important first step in arguing an environmental ethics that includes respect for plant life.

Areas of concern: wild nature, cultivated nature, utilized nature

Persons interfere with plant life in many ways. In the age of climate change and systemic ecological crises, the anthropogenic effects on vegetative nature are even deeper and more comprehensive than they used to be. In order to make plant ethics concrete, it is helpful to divide attention to plant life according to what spheres of the human–nature relationship they are encountered in.

Following Taylor (1986), nature can be divided into different spheres according to different relationships between humans and nature. The first sphere is *wild nature*, which is nature largely untouched by intentional human designs and processes. Even though wild nature may not exist in the literal sense – as significant and long-lasting anthropogenic influence is at work in all corners of the planet – there still are areas of nature that are not excessively disturbed by human infrastructures and development, some of these areas have been given the status of "national parks". Wild nature in this non-literal sense also exists in cities and in the midst of cultivated nature. In all these cases, respect for the flourishing of plants claims leaving nature alone. If we want plant life to keep on unfolding evolutionary processes undisturbed by humans, we need to leave and make space for "wild" nature. Plants need space to which they have been adapted and which they can use for developing new species.[3]

The second sphere is that of *cultivated nature*. Part of what makes for a civilization is precisely this cultivation of nature, and particularly the breeding of plants. Cultivated nature includes landscapes shaped by farming, horticulture, and agriculture. Respect for the flourishing of plants in this second sphere demands, among other things, that cultivation techniques are chosen so that plants can develop their typical characteristics and fulfill their life-cycle. In this second sphere, however, plant ethics expands its concerns beyond the flourishing of individual plants, as the cultivation of nature is the collective enterprise to shape natural resources and landscapes. Respect for flourishing becomes here part of a much broader assessment of the value of plant life, extending, for example, also to issues in water ethics and sustainable land use.[4]

The third sphere is that of *utilized nature*: nature that is transformed and used so that its most basic life-processes are shaped and modeled by humans and in view of human objectives.[5] Nature is utilized when cells are modified, in smart farming and when working with biofacts (Karafyllis 2001). When utilizing nature, plant ethics can at least shift the burdens of proof when judging technologies and methods of breeding. Note that to say that plants have the capacity to flourish does not necessarily mean that each single plant should flourish. It does mean, however, that technologies which have a deep impact on plants' capacities to flourish need to undergo a particularly thorough scrutiny. I have argued elsewhere (Kallhoff 2009) that technologies which prevent plants from generating seed, as for example the terminator gene, should be criticized as mistreatment of plant life.

When it comes to utilized nature, it is difficult to judge whether controlling plant life through smart farming, or changing the most basic characteristic of plants by means of genetic engineering, is right or not. Ethical deliberation in these cases will include principles of precaution, but also debates about sustainability. Here again, plant ethics wants manipulation to be justified: technologically innovative breeding technologies are needed in order to enhance the food supply, but they still need to be evaluated in reference to a number of criteria, including their effects on plant flourishing. Against the backdrop of an ethics of plant flourishing there is also a difference between using plant tissue or cells in order to develop new materials or medicines, and designing plants that can no longer flourish on their own. Unlike the former, the latter technologies undermine that which plants essentially are: a life-form that strives to develop a good life.

Conclusion

In this chapter, I have argued for an interpretation of plant life that mediates between two opposite ideas. Plants are neither presented as having moral standing because of some plant-self that claims respect; nor are plants regarded as entities that are morally comparable to stones that can be kicked around. Instead, the study of plant life in botany tells us that plants strive to realize a life that is good in their own terms. They strive to flourish.

As for the moral conclusions that have been drawn from this interpretation of plant life, I have argued for a cautious and multi-faceted way to argue plant ethics – yet one that takes flourishing as its baseline. Even though plants do not suffer in a way comparable to the way that animals suffer, they can objectively be harmed or benefited. Following an analysis of von Wright, the categories of "the beneficial" and "the harmful" have a causal and an axiological component. Both can be applied to plant life. Following this assessment, it is also right to claim respect for flourishing, yet without claiming that it is the plant itself that matters. Instead, the anchor that an approach to flourishing provides for a moral theory is the distinction between constructive behavior as opposed to destructive behavior regarding plant life. I have also argued that three spheres of human–plant relationship need to be distinguished. The distinction helps us translate the rather abstract claim to respect plant life into principles of adequate treatment of plants in different contexts.

Since life on Earth is competitive, it is important not only to respect a variety of life-forms; it is also important to discuss cases in which decisions that are harmful to some and that are good to others need to be made. Overall, a neo-Aristotelian approach seeks to argue for win–win scenarios whenever this is possible. In particular, it draws attention to sensible categories in plant life, including water supply, soil, and space more generally. The enhancement of conditions of plant flourishing is not only a sound claim in a plant ethics. It is also part of a more comprehensive approach to an ethical framework that takes the living-conditions of various types of beings seriously. If humankind wishes to continue to live a good life on planet Earth, it is important to respect plant life.

Notes

1. I do not imply the concept of a "species", because this concept is not only highly controversial, but also rather a rough guide to the cartography of the diversity of life-forms. For the concept of "flourishing" it suffices to call attention to inborn capacities that are expressed in typical ways by comparable living beings. This approach fits the proposal of Thompson to speak of a "life form" in terms of specific and particular ways to realize shared capacities which are typical of groups of living beings (see Thompson 1995).
2. This sentence is of course meant to provide an analogy to Thomas Nagel's famous article on animal life. He argues that we persons simply cannot know how it is like to be a distinct animal (Nagel 1974).
3. Therefore, "ecological space" is particularly important in plant life. The category of "ecological space" and its moral dimensions have recently been explored in (Hayward

2017). Hayward regards ecological space as morally important for human beings, but I believe the notion can usefully be extended to plants, as the most basic condition for the life of individual plants to unfold and for plant species to evolve.
4. A proposal for a value-based assessment of the design of water-resources that also includes ideas about fairness to nature is in Kallhoff 2014b.
5. Note that this sphere differs from Taylor's (1986) proposal.

References

Affifi, R. 2013. Learning plants: Semiosis between the parts and the whole. *Biosemiotics* 6, no. 3: 547–59.

Attfield, R. 1981. The good of trees. *Journal of Value Inquiry* 15, no. 1: 34–54.

Böhme, G. 1989. *Für eine ökologische Naturästhetik* (1st ed.). Frankfurt am Main: Suhrkamp.

Cooper, J. M. 1986. *Reason and human good in Aristotle*. Indianapolis, IN: Hackett Pub. Co.

Frankena, W. K. 1979. *Ethics*. Englewood Cliffs, NJ: Prentice-Hall.

Hayward, T. 2017. Ecological space. The concept and its ethical significance. Pp. 311–20 in *The Oxford handbook of environmental ethics*, edited by S. M. Gardiner and A. Thompson. Oxford; New York: Oxford University Press.

Heil, M. and Karban, R. 2010. Explaining evolution of plant communication by airborne signals. *Trends in Ecology & Evolution* 25, no. 3:137–44.

Hirt, H., ed. 2009. *Plant stress biology: From genomics to systems biology*. Weinheim: Wiley-Blackwell.

Johnson, D. and Gilbert, L. 2015. Interplant signalling through hyphal networks. *New Phytologist* 205, no. 4: 1448–53.

Kallhoff, A. 2002. *Prinzipien der Pflanzenethik. Die Bewertung pflanzlichen Lebens in Biologie und Philosophie*. Frankfurt; New York: Campus Verlag.

Kallhoff, A. 2009. Das Terminatorgen aus pflanzenethischer Perspektive. *Journal für Verbraucherschutz und Lebensmittelsicherheit* 4, no. 3: 361–65.

Kallhoff, A. 2014a. Plants in ethics: Why flourishing deserves moral respect. *Environmental Values* 23, no. 6: 685–700.

Kallhoff, A. 2014b. Water justice: A multilayer term and its role in cooperation. *Analyse & Kritik* 36, no. 2: 367–82.

Karafyllis, N. C. 2001. *Biologisch, Natürlich, Nachhaltig. Philosophische Aspekte des Naturzugangs im 21. Jahrhundert*. Tübingen: Francke.

Larcher, W. 2001. *Ökophysiologie der Pflanzen: Leben, Leistung und Streßbewältigung der Pflanzen in ihrer Umwelt* (6th revised ed.). Stuttgart: Ulmer.

Mancuso, S., and Viola, A. 2015. *Brilliant green. The surprising history and science of plant intelligence*. Washington; Covelo; London: Island Press.

Nagel, T. 1974. What is it like to be a bat? *The Philosophical Review* 83, no. 4: 435–50.

Sorabji, R. 1975. Body and soul in Aristotle. Pp. 42–64 in *Articles in Aristotle, Vol. 1*, edited by R. Sorabji, M. Schofield and J. Barnes. London: Duckworth.

Taylor, P. 1986. *Respect for nature: A theory of environmental ethics*. Princeton, NJ: Princeton University Press.

Thompson, M. 1995. The representation of life. Pp. 247–96 in *Virtues and reasons. Philippa foot and moral theory*, edited by R. Hursthouse, G. Lawrence and W. Quinn. Oxford: Clarendon Press.

Wright, G. H. von. 1972. *The varieties of goodness*. London: Routledge and Kegan Paul.

5 The dignity of plants

An overview of the discussion in German-speaking countries

Sabine Odparlik

Introduction

The concept of the dignity of creatures (*Würde der Kreatur*)[1] entered the Federal Constitution of the Swiss Confederation (*Bundesverfassung der Schweizerischen Eidgenossenschaft, BV*) in May 1992 via referendum. The relevant article governs the protection of human beings and their environment against the misuse of gene and reproduction technology:

> The Confederation shall legislate on the use of reproductive and genetic material from animals, plants and other organisms. In doing so, it shall take account of the dignity of living beings as well as the safety of human beings, animals and the environment, and shall protect the genetic diversity of animal and plant species.
>
> (Swiss Confederation 1999, Art. 120 § 2 BV)

According to Baranzke (2001) this wording reflects different, partially competing views regarding how human beings should act with respect to the non-human world. Baranzke distinguishes the following three positions, which were in evidence in discussions on the constitutional amendment before the referendum: an anthropocentric position, focused on the protection of human beings; a holistic position, aiming at the protection of all creatures; and a third position emphasizing animal protection and welfare (Baranzke 2001). All of these positions interpret the notion of the dignity of creatures in their own ways and for their own reasons. As Brom correctly noted, this has "raised a series of questions that need to be answered in the Swiss legislative process" (Brom 2000).

An important question regards the scope of the notion of dignity of creatures. Experts invited by the *Swiss Agency for the Environment, Forests and Landscape*[2] came to the conclusion that at least animals and plants can have dignity (Balzer, Rippe, and Schaber 1998; Praetorius and Mueller 1996). The issue remains open, however, and the Federal Ethics Committee on Non-Human Biotechnology (ECNH), founded in 1998, states that its "key tasks include putting this concept into concrete terms" (Federal Ethics Committee on Non-Human Biotechnology [ECNH] 2017).

The following two statements by the ECNH are particularly noteworthy: "The Dignity of Animals" (ECNH/SCAE 2001) in cooperation with the Swiss Committee on Animal Experiments (SCAE), and "The Dignity of Living Beings with Regard to Plants" (ECNH 2008). In comparing these booklets, it is striking that it took much longer to prepare the reports on plants. Additionally, the titles of the two statements illustrate an increased amount of cautiousness on the side of the ECNH when applying the notion of dignity to plants. The contents of these two publications also show other significant differences: whereas in the case of animals the committee succeeded in drawing up a proposal for the systematic evaluation of damage to dignity, in the case of plants the ECNH had to admit that the "only criterion on which all the members could agree [. . .] was that we should not harm or destroy plants arbitrarily. Whether concrete ways of acting could be derived from this prohibition on the arbitrary handling of plants, and what those ways of acting might be, remained unclear" (ECNH 2008). The ethical debate about appropriate human behavior towards non-human beings, triggered by the Swiss constitutional amendment of 1992, painted a similar picture (Kunzmann 2007): whereas the notion of the Dignity of Animals became very popular and was widely discussed in German-speaking countries, only few devoted much attention to the Dignity of Plants.

The discussion of the Dignity of Plants is characterized by great restraint and caution. Publications addressing this concept are usually met with harsh criticism or even ridicule. A vivid example is the awarding of the 2008 Ignoble Peace Price to the ECNH for its publication on Plant Dignity. These negative responses might be based on a mere cultural/linguistic misunderstanding. There is a significant divergence in meaning between the term "dignity" and the German term "Würde." The origin and etymology of the latter suggest that the German concept originated from the term "*werden*" (English: to become, to evolve), which refers to any directed development (Lühr 2010) and is thus better suited for application to non-human living beings than is the concept of dignity, historically designating a high status attributed to Roman nobility and later to all humans (O'Malley 2010). It seems difficult, for example, to ascribe dignity to plants in French: in 1999, the French version of the BV, the term "dignité de la creature" has been changed to "intégrité des organismes vivants."

But there are different meanings of the term "*Würde*" within the German language as well. Indeed, the most widely accepted conception of dignity denotes the unique moral status of all human beings (*Menschenwürde*). As Ingensiep (2001) has argued, this conception is importantly related to some core ethical ideas that are characteristic of Western culture, including the idea that humans are at the top of the *Scala Naturae*, the Great Chain of Being, followed by animals, plants, and minerals. It is hard to transfer the notion of dignity to plants when the dignity of human beings, to whom the notion of dignity is typically applied, depends in part precisely on their being different from and superior to plants. To this day, it is widely held among German-speaking philosophers, as well as the public, that plants are only indirectly morally considerable: insofar as they are appreciated or needed by humans (and possibly animals).

Background

Since the 1980s, the perception of plants in public discourse has slowly shifted from machine-like creatures to real living beings with specific needs. This has drawn plants closer to direct moral considerability. The following aspects have induced this change, and the discussion about the Dignity of Plants may be only one of its indicators:

1. Environmental ideas reached public consciousness after the 1960s, when changes in the natural environment due to human influence began to be understood as an environmental crisis. Supported by scientific findings, these changes made the interdependence of all living beings obvious. This gave rise to environmental movements, and it was indeed the environmental movement in Switzerland that prepared the ground for the referendum on the dignity of creatures (Praetorius and Mueller 1996).
2. The ideas of environmental philosophers from English-speaking countries were also influential. In search of an appropriate ethical response to ensure the preservation or restoration of degraded environments and their components, environmental ethicists cannot exclude plants from their considerations. Attfield (1981) and Taylor (1986), who argued for the moral consideration of all living beings, significantly influenced the discussion about the dignity of living beings, including plants, in German-speaking countries.
3. Increased scientific knowledge of plant physiology indicates that differences between plants and animals are less sharp than has been assumed so far. Concepts of intelligence, individual behavior, intention, learning, communication and memory are now applied to plants by some scientists (Koechlin 2014, Mancuso/Viola 2015; Trewavas 2003, 2004). Though suspiciously eyed by other biologists, a new branch of plant physiology, called plant neurobiology, has emerged (Alpi et al. 2007).[3]
4. New biotechnological methods, most notably genetic engineering, have blurred the lines between different species by providing methods to introduce foreign or synthetic genetic material into an organism of interest, and by producing chimeras and hybrids; but they have also illustrated the common origins of all living beings. The latter was one of the considerations encouraging the Swiss to widen the scope of protection against the abuse of reproductive and gene technology from humans to all living beings (Federal Council of the Swiss Federation 1998). The aspect of common origin finds recognition also in the choice of a notion like Dignity of Creatures, though within a theological interpretation (Baranzke 2001).
5. As Verhoog (2001, 15–16) notes: "the concept of animal integrity is [. . .] brought in to address those infringements of the animal's nature which go beyond health and wellbeing." For him, it is a "logical consequence of the introduction of this concept [. . .] that, because it is independent of sentiency, it can be applied to all animals." Against this background, he asks whether "the same way of reasoning does not automatically lead to the idea that plants also have a nature of their own [. . .], an integrity which can be

> violated." Much like integrity, the notion of dignity also goes beyond pain and injury. If dignity is emphasized, the exclusion of plants from the circle of the morally considerable is no longer convincing if only supported by the argument that plants are not capable of pain and suffering, and that humans cannot be cruel to plants.

Of course, questions regarding the moral and legal status of plants have arisen also in non-German-speaking countries. However, these questions have been discussed in different terms – using concepts like autonomy, inherent or intrinsic value, or integrity of plants. The notion of dignity seems to be peculiar to the German-speaking world. What could have motivated advocates of the moral and legal status of plants to choose that term over others?

Meaning

Supporters of the Dignity of Plants propose various interpretations of the concept. However, they all converge on at least the following point: plants have to be morally considered for their own sake, rather than because we use them and they please us. According to this non-anthropocentric position, the dignity of plants can be violated by excessive human use.

Here, the dignity of plants shares grounds with the dignity of human beings, but with one crucial difference. It is a widely held view that human beings should be protected from degradation in all circumstances. This implies that any and all violations of human dignity are morally impermissible, and this in turn excludes trade-offs between the dignity of one person and other values and interests, and even between the dignity of two persons. But violations of plant dignity can be morally permissible, even in the eyes of its proponents, as long as they are non-arbitrary and result from a considerate balancing of morally relevant interests and the needs of humans and plants (Odparlik 2010).

Why dignity?

The application of dignity to non-human living beings may appear odd to non-German speakers, especially when even human dignity is not a central part of the corresponding public discussion. However, even when this is the case, in light of the different consequences of the two concepts of dignity, the question remains unanswered as to why terms already established in this context, such as integrity, inherent or intrinsic value, could not replace the notion of dignity of non-human living beings without a loss, at least for German speakers (Odparlik 2013). Several aspects might have led to the application of the concept of dignity to plants – as follows:

- First, the concept of dignity, unlike these other concepts, is already charged with a recognition of moral status and an implicit demand for direct moral consideration (Hauskeller 2003; Siegetsleitner 2007; Sitter-Liver 1999, 2008).

- Second, the dignity of plants inherits, at least in part, the rhetoric peremptoriness of moral commands regarding the dignity of humans, which are held to be of particularly high significance within morality, ethics, law, politics and in public discussion. This high profile of the notion of dignity appears to make it particularly suitable to a public discussion about moral standards for relating to plants: dignity is both a powerful moral standard and a catchy term that resonates strongly within most of us.
- Third, the notion of dignity often figures in our discourses as a critical benchmark for evaluating the quality of the treatment that individuals receive from others or from governments. Employing the notion of dignity thus alerts us to the fact that the fate of individuals is at stake in a given situation. This *prima facie* connotation of individuality of the term "dignity" is not shared by the terms "integrity" or "intrinsic value." Here is worth noting, however, that such focus on individuals is applicable to plants but tends to discourage more holistic ethical views that see plants as populations and/or elements of complex eco-systems (and valuable and morally considerable as such). This is part of a larger discussion at the intersection of plant physiology and philosophy about whether plants are to be thought of as individuals at all, and such doubts have led some advocates of the dignity of non-human beings to restrict the application of the concept to animals (Arz de Falco 2001; Balzer/Rippe/Schaber 1998; ECNH 2008; Kunzmann 2007; Odparlik 2008; Sitter-Liver 2008).

Conceptual premises

Autonomy

Some earlier interpretations of the dignity of non-human beings, including plants, are strongly indebted to the Kantian tradition. In Kant's moral theory, the foundation of human dignity is autonomy. Kantian autonomy is shown by a person who acts according to laws rationally laid down by him- or herself. Autonomous beings are moral agents. According to Kant, every moral agent is an end in him- or herself, and thus assigned with dignity – whereas all other creatures that are not moral agents are assigned with no dignity, and can be used merely as means (Kant 1785/2002). Frequently, it is suggested that anthropocentric approaches follow Kant in this point, and therefore are in line with the Western tradition of assumed human superiority in claiming that humans, unlike all other living beings, have to be considered for their own sake (Hurst and Mauron 2012).

Some advocates of the Dignity of Plants challenge this human-centered position arguing that not only humans and animals, but also plants can be characterized as autonomous beings and therefore can be considered as ends in themselves. Some biologists, for example, describe plants as intelligent beings (Hall 2011; Mancuso and Viola 2015), while others argue that plants are agents at least in the minimal sense that they pursue an end that is their own good as they attempt to maximize their survival, growth, and reproduction (Praetorius and Mueller 1996; Sitter-Liver 1999).

Used as grounds for establishing the dignity of plants, this idea of plants as autonomous beings generates not only theoretical but also practical puzzles: "the assumption of a substantial equality of the dignity concepts for humans and for animals or plants would cause a very far-reaching protection of all life forms in a manner similar to the protection of human life" (Schmidt 2001). This implication might look implausible to many. To avoid it, plant dignity proponents are ready to admit that the dignity of plants, unlike that of humans, can sometimes be justifiably violated. Critics, at this point, have an easy go at pointing at the inconsistency.

More fundamentally, the application of the Kantian notion of autonomy to creatures that are not capable of acting as moral agents is described by critics as a massive philosophical misunderstanding (Baranzke 2008; Knoepffler 2011). Autonomy in the Kantian sense has to do with moral purposes, and these are not in the realm of nature that plants inhabit. There, only natural purposes exist; and these are better discussed in an Aristotelian language (Odparlik 2013).

Own good

Many advocates of Plant Dignity accept the above objection and have turned Aristotelian. Following Paul Taylor's (1986) biocentric theses in *Respect for Nature*, they stick to a teleological interpretation of the development and function of living beings. The concept of plant teleology is philosophically controversial, but this is true also of the concepts of human autonomy and moral agency. However, Taylor's Aristotelian suggestion that we conceive of each individual organism "as a teleological center of life, pursuing its own good in its own way" (Taylor 2005, 76) seems to support the concept of Plant Dignity insofar as an individual plant's own good is its *telos*, and this provides a non-anthropocentric criterion for the evaluation of human dealings with plants. The idea that every living being – conscious or not – pursues its individual good provides a suitable foundation for the moral claim that in light of their dignity, plants have to be considered for their own sake and not for the interests and needs of humans, animals, or even nature as a whole.

Integrity

Once plants are said to deserve moral consideration in light of their dignity, and the dignity itself is defined with respect to their own good, we still need an explanation as to how the *telos* of a certain plant can be frustrated. To approach this question, the concept of a plant's own good has to be clarified. As Teutsch (1995) and Lötscher (2000) note, the notion of integrity can be useful in this context. However, there are also multiple conceptions of integrity (Hauskeller 2007). Two of those conceptions are particularly noteworthy, and they can be thought to point to two different positions on the Dignity of Plants – which I shall call "conservative" and "liberal" positions.

Conservative positions propose an interpretation of the integrity of living beings that "refers to their nature or way of being, their wholeness, completeness, their species-specific characteristics" (Lammerts van Bueren et al. 2003, 1923). In this sense, the integrity of an organism, as a certain state of phenotype and genotype to be conserved, is equated with its own good. From this point of view, any distortion of a plant's *telos*, any change to a plant's species-specific properties, parts, or functions due to human action, is considered "unnatural" and thus a violation of that plant's dignity. On these grounds, it has been argued that genetic engineering of plants, in contrast to traditional breeding techniques, always assault the dignity of creatures because they entail the transgression of species boundaries (Praetorius and Mueller 1996).

It is, of course, an open question whether moral judgments and precepts can be derived from nature at all. It is also unclear what counts for a "natural" state. In addition, however, the conservative position on the Dignity of Plants is troubled by questions that are distinctively its own. For instance, it is unclear which properties, parts and functions of an individual plant are species-specific. Balzer et al. (1998) suggest asking biologists, but as Ereshefsky notes, biologists and philosophers disagree on the definition of the term "species." Moreover, even if agreement was found on this question, it is unlikely that we'll find biological traits that occur in each, all, and only the members of a certain species (Ereshefsky 2010). Any given selection of certain traits whose protection is essential to protect Plant Dignity can easily be criticized as an expression of human arbitrariness, whether or not it is in accord with the recommendations of biologists. This way of providing moral consideration to plants is in some way in conflict with the Dignity of Plants, as it turns out to obey human standards (including scientific standards as used by biologists) rather than be devoted to considering a plant for its own sake (Odparlik 2010).

Liberal positions (Odparlik 2011) try to circumvent the problems associated with the definition of an individual's own good by referring to a different conception of integrity, which is well characterized by the term "flourishing." This interpretation does not emphasize the conservation of the species-specific characteristics of an organism. On this view, respecting plant dignity means considering the ability of plants to adapt to environmental conditions and therefore to survive, grow, develop, and thrive (Kallhoff 2002). Here, the integrity of a plant is not identified with its own good understood as a certain state of an organism, but rather understood as an essential precondition for the development of a plant toward its *telos*. The realization of such *telos* is only possible for a plant if it is able to constantly integrate its components and functions by coordinating their interaction and cooperation. A plant's ability to flourish should be the focus of our moral concern, not a plant's own good, which might be quite difficult to define for the reasons mentioned in the previous section.

This less common position is called liberal because, by its lights, changing species-specific characteristics can be morally permissible as long as a plant is not hindered in preserving its integrity. Hence, the production of transgenic plants, for example, is not thought to be morally impermissible in every case, but techniques that undermine the capacity of plants to survive, grow, adapt, and thrive – like the so-called terminator technology – or hinder embryogenesis within the seeds of transgenic plants, are morally criticized (Kallhoff, 2014).

Further debate

The notion of the Dignity of Plants emerged, in German-speaking countries, mostly in the context of a dispute on the moral permissibility of genetic engineering. Thereupon it became an issue not only of public but also of philosophical concern. Even though the application of the term "dignity" to plants has been met with severe criticism, it has also had the effect of fostering a debate on how plant ethics should be theoretically construed and then implemented in our moral practice. No conclusive argument has been provided to the effect that we ought to ascribe dignity to plants, and therefore to oblige ourselves to according direct moral consideration to plants. One reason for that may be the assumed interdependency of the meaning of the term "Dignity of Plants" with the controversial thesis of the teleological constitution of all living beings.

What normative implications follow from the recognition of the dignity of plants? To put it directly: how should we handle plants? Two main proposals have been advanced, which come to different results regarding, for example, the assessment of plant genetic engineering. One aims at the conservation of the plant's own good as a certain state of a being according the typical characteristics of its species; the other, which I have called a liberal position, strives for the preservation of the ability of plants to survive, grow, adapt and thrive, and therefore on the precondition of their development towards their own good. A decision on this issue must urgently be taken, particularly in countries, such as Switzerland, which have inserted in their constitutions the dignity of creatures and must now make policy regarding the genetic material of plants and human interventions on it.

Nevertheless, one result of the debate about consequences can already be acknowledged, namely the fact that both, the techniques of genetic engineering and the common handling of plants, require moral reasoning. The Dignity of Plants, as a moral principle, cannot be limited to questions regarding genetic engineering.

Concerns also arise regarding the risk of imposing excessive moral demands on people by ascribing dignity to plants. These concerns miss a crucial point, however, which I have mentioned several times in this chapter. Most advocates of Plant Dignity insist on a difference between human dignity and the dignity of non-human living beings: only the dignity of human beings ought to be protected in any case and insulated from all trade-offs. When morally considering plants on grounds of their dignity, on the other hand, trade-offs are not ruled out. The main question that we find ourselves confronting in the case of plants is not whether and when we are allowed to use plants, but rather under what circumstances should humans be willing to subordinate their own interests to the dignity of plants. To answer this question, we need to decide when the violation of a plant's dignity is so severe that it cannot be morally justified (or can only be justified in extreme circumstances). This, in turn, will force us to clarify what sort of moral beings we, humans, intend to be. Taking the dignity of plants into account, then, will push us to a reconsideration of ourselves that goes well beyond the realm of plant ethics.

Notes

1. Suggested alternative translations of the constitutional term "Würde der Kreatur" are "Dignity of Creation" and "Dignity of Living Beings".
2. Since 2006 the Federal Office for the Environment.
3. Influenced by these new studies, a group of Swiss, Austrian, and German authors has laid down a set of rights for plants: see the *Rheinauer Thesen zu Rechten von Pflanzen* 2016. Accessed 15 January 2018 from www.blauen-institut.ch/s2_blue/tx_blu/tp/tpt/t_rheinau.pdf.

References

Alpi, A., Amrhein, N., Bertl, A., Blatt, M. R., Blumwald, E., Cervone, F., Dainty, J., De Michelis, M. I., Epstein, E., Galston, A. W., Goldsmith, M. H., Hawes, C., Hell, R., Hetherington, A., Hofte, H., Juergens, G., Leaver, C. J., Moroni, A., Murphy, A., Oparka, K., Perata, P., Quader, H., Rausch, T., Ritzenthaler, C., Rivetta, A., Robinson, D. G., Sanders, D., Scheres, B., Schumacher, K., Sentenac, H., Slayman, C. L., Soave, C., Somerville, C., Taiz, L., Thiel G. and Wagner, R. 2007. Plant neurobiology: No brain, no gain? *Trends in Plant Science* 12, no. 4: 135–36. DOI: 10.1016/j.tplants.2007.03.002.

Arz de Falco, A. and Müller D. 2001. *Wert und Würde von "niederen" Tieren und Pflanzen. Ethische Überlegungen zum Verfassungsprinzip "Würde der Kreatur"*. Fribourg: Paulusverlag.

Attfield, R. 1981. The good of trees. *Journal of Value Inquiry* 15, no. 1: 35–54. DOI: 10.1007/BF00136626.

Balzer, P., Rippe, K. P. and Schaber, P. 1998. *Menschenwürde vs. Würde der Kreatur*. Freiburg: Alber.

Baranzke, H. 2001. *Würde der Kreatur?* Würzburg: Königshausen & Neumann Press.

Baranzke, H. 2008. "Wozu brauchen Pflanzen Würde? Eine Anfrage aus Kantischer Perspektive." Pp. 36–60 in *Wie die Würde gedeiht. Pflanzen in der Bioethik*, edited by S. Odparlik, P. Kunzmann and N. Knoepffler. München: Utz.

Birnbacher, D. 2013. "Warum hat der Mensch weitergehende Rechte als andere Naturwesen?" Pp. 443–463 in *Autonomie und Würde. Leitprinzipien in Bioethik und Medizinrecht*, edited by H. Baranzke and G. Duttge. Würzburg: Königshausen & Neumann.

Brom, F. W. A. 2000. The good life of creatures with dignity: some comments on the Swiss expert opinion. *Journal of Agricultural & Environmental Ethics* 13, no. 1–2: 53–63. DOI: 10.1007/BF02694135.

ECNH. 2000. Stellungnahme zur französischen Version des Art. 120 BV. Accessed 13 July 2016 from www.ekah.admin.ch/fileadmin/_migrated/content_uploads/d-Stellungnahme-FrVers-Art129BV-2000_03.pdf.

ECNH. 2008. *The dignity of living beings with regard to plants*. Bern: Ackermanndruck.

ECNH/SCAE. 2001. Die Würde des Tieres, Schaub & Rüedi, Bern.

Editorial. 2008. Open to interpretation. *Nature* 453, 824. DOI: 10.1038/453824b.

Ereshefsky, M. 2010. Species. In *The Stanford Encyclopedia of Philosophy* (Summer 2016 edition), edited by E. N. Zalta. Accessed 13 June 2016 from http://plato.stanford.edu/archives/sum2016/entries/species/

Federal Council of the Swiss Federation. 1998. Dispatch of the Federal Council of the Swiss Federation for the Federal Assembly from 18th of September 1998. BBI 1989 III.

Federal Ethics Committee on Non-Human Biotechnology (ECNH). 2017. Dignity of living beings. Accessed 24 January 2018 from http://www.ekah.admin.ch/en/topics/dignity-of-living-beings/

Hall, M. 2008. *Plants as persons: A philosophica botany*. Albany, GA: State University of New York Press.

Hauskeller, M. 2003. Integrity and dignity. Is there a difference? *Revista Romana de Bioethica* 1 (4):81–88.

Hauskeller, M. 2007. *Biotechnology and the Integrity of Life*. Aldershot: Ashgate Publishing.

Hurst, S. and Mauron, A. 2012. Expérimentation animale: une peseé en déséquilibre *Bioethica Forum* 5 (1): 20–22.

Ingensiep H. W. 2001. *Geschichte der Pflanzenseele*. Stuttgart: Alfred Kröner Press.

Kallhoff, A. 2002. *Prinzipien der Pflanzenethik: Die Bewertung pflanzlichen Lebens in Biologie und Philosophie*. Frankfurt am Main: Campus Verlag.

Kallhoff, A. 2014. Plants in Ethics: Why Flourishing Deserves Moral Respect. *Environmental Values* 23 (6): 685–700.

Kant, I. (1785) 2002. *Groundwork for the Metaphysics of Morals*, translated by. A. Zweig, edited by E. Hill Jr. and A. Zweig. Oxford, UK: Oxford University Press.

Koechlin, F. (ed.). 2014. *Jenseits der Blattränder. Eine Annäherung an Pflanzen*. Basel: Lenos.

Koechlin, F. 2007. Die Würde des Eichenblatts. Pp. 57–72 in *Eine Würde für alle Lebewesen?* edited by S. Odparlik and P. Kunzmann. München: Herbert Utz Press.

Koechlin, F., Ammann, D., Gelinsky, E., Haerlin, B., Ott, M., Sitter-Liver, B., Stumpf, W., Wagne, E., and Zschunke, A. 2016. *Rheinauer Thesen zu Rechten von Pflanzen*. Accessed 13 June 2016 from www.blauen-institut.ch/s2_blue/tx_blu/tp/tpt/t_rheinau.pdf

Kunzmann, P. 2007. *Die Würde des Tieres*. Freiburg: Alber.

Lammerts van Bueren, E. T., Struik, P. C., Tiemens-Hulschera, M. and Jacobsen, E. 2003. Concepts of intrinsic value and integrity of plants, in organic plant breeding and propagation. *Journal for Agronomy & Crop Science* 43: 1922–29.

Lötscher, A. 2000. Würde der Kreatur – integrité des organismes vivants. *LEGES* 2: 137–56.

Lühr, R. 2010. Wertbegriffe in indogermanischen Sprachen Europas. Pp. 253–76 in *Die Menschenwürde im internationalen Vergleich*, edited by C. Baumbach and P. Kunzmann. München: Herbert Utz Press.

Mancuso, S., Viola, A. 2015. *Die Intelligenz der Pflanzen*. München: Zentrum Verlagsgesell-schaft.

O'Malley, M. 2010. Dignity in US-Bioethics debate: Needs Wuerde. Pp. 253–76 in *Die Menschenwürde im internationalen Vergleich*, edited by C. Baumbach and P. Kunzmann. München: Herbert Utz Press.

Odparlik, C. f. S. 2008. Die Individualität von Pflanzen im Kontext der Diskussion um die Würde der Kreatur. Pp. 275–300 in *Wie die Würde gedeiht. Pflanzen in der Bioethik*, edited by S. Odparlik, P. Kunzmann and N. Knoepffler. München: Utz.

Odparlik, C. f. S. 2010. *Die Würde der Pflanze*. Freiburg: Karl Alber Press.

Odparlik, C. f. S. 2013. Würde – ein angemessener Begriff in Bezug auf nichtmenschliche Lebewesen? Pp. 507–28 in *Autonomie und Würde. Leitprinzipien Bioethik und Medizinrecht*, edited by H. Baranzke and G. Duttge. Würzburg: Königshausen & Neumann.

Odparlik, E. G. S. 2011. Die Würde nichtmenschlicher Lebewesen – jenseits des bio-konservativen Sprachspiels. Pp. 31–42 in *Gentechnik – Pflanzen, Tiere und das Humane*, edited by P. Kunzmann, S. Odparlik, and N. Knoepffler, Stuttgart: Königshausen & Neumann.

Praetorius, I. and Mueller, D. 1996. *Die Würde der Kreatur*. Schriftenreihe Umwelt No. 260. Bern.

Schmidt, H. 2001. "The dignity of man and the intrinsic value of the creature." Pp. 19–23 in *Intrinsic value of plants in the context of genetic engineering*, Ifgene Workshop Proceedings 9–11 May, edited by D. Heaf and J. Wirtz. Hafan: Ifgene.

Siegetsleitner, A. 2007. Zur Würde nichtmenschlicher Lebewesen. Pp. 103–18 in *Eine Würde für alle Lebewesen?* edited by S. Odparlik and P. Kunzmann. München: Utz.

Sitter-Liver, B. 2008. Würde als Grenzbegriff. Erläutert am Beispiel der Würde der Pflanze. Pp. 171–82 in *Wie die Würde gedeiht. Pflanzen in der Bioethik*, edited by S. Odparlik, P. Kunzmann, and N. Knoepffler. München: Utz.

Sitter-Liver, E. G. B. 1999. *Würde der Kreatur. Eine Metapher als Ausdruck anerkannter Verpflichtungen*. Pp. 455–78 in *Philosophisches Jahrbuch 106*, edited by H. M. Baumgartner, J. Jacobi, H. Ottmann et al. München: im Auftrag der Görres Gesellschaft.

Swiss Confederation. 1999. *Federal Constitution of the Swiss Confederation*. Art. 120 § 2 BV, pp. 2–80.

Taylor, P. W. 2005. Respect for nature. Pp. 74–84 in *Environmental ethics*, 5th ed., edited by A. Light and H. Rolston III. Malden, Oxford & Carlton: Blackwell Publishing Press.

Teutsch, G. M. 1995. *Die "Würde der Kreatur". Erläuterungen zu einem neuen Verfassungs-begriff am Beispiel des Tieres*. Stuttgart: Haupt.

Trewavas, A. 2003. Aspects of plant intelligence. *Annals of Botany* 92, no. 1: 1–20. DOI: 10.1093/aob/mcg101.

Trewavas, A. 2004. Aspects of plant intelligence: An answer to Firn. *Annals of Botany* 93, no. 4: 353–57. DOI: 10.1093/aob/mch059.

Verhoog, H. 2001. The intrinsic value of animals: Its implementation in governmental regulations in the Netherlands and its implication for plants. Pp. 15–16 in *Intrinsic value of plants in the context of genetic engineering, Ifgene Workshop Proceedings* 9–11 May, Hafan, edited by D. Heaf, and J. Wirtz.

6 Facing only outwards?

Plant bodily morphogenesis and ethical conceptual genesis

Karen Houle

Plant ethics: Angles of approach

Why "plant ethics"? There are many good reasons for thinking ethically about plants. Plants have immense instrumental value. Plants are so crucially important to the health and vitality of all other living beings on this planet that developing a theoretically sound protectorate of the vegetal kingdom and deriving a policy of conservation on their behalf by way of that protectorate is prudent. However, if we only care about plants (trees for instance) because they perform the vital service of making oxygen, then the protection need not specify anything more precise than safeguarding a minimum amount of CO_2-fixing biomass. The *astonishing multiplicity* (Derrida 2008, 48) that is the actual plant kingdom is entirely bypassed by such an approach,[1] as is the concept of their intrinsic value. One alternative approach to plant ethics (which I do not defend here) could get traction by pointing out that *if* we are prepared to accept the premise that animals – polar bears, tuna, newts, etc. – have intrinsic value, and thus that we have ethical duties towards them in regards to that value, *then* it would be inconsistent to withhold such a premise with respect to plants: orchids, sequoias, tall grass prairies, and so on.

Another intelligible angle of approach for the topic of plant ethics is to explore the possibility that individual plant subjects – particularly such trees as the stalwart *Acer saccarum* across the road from where I am writing this paper, or even the less impressive potted jade plant on my desk – can be the object of application of some vegetal version of the important humanist ethical concepts that we employ when we try to extend and ground moral considerability to entities other than humans. The ideas that plants have a kind of *autonomy* (Hall 2009, 169–181), or *dignity* and *integrity* (Federal Ethics Committee on Non-Human Biotechnology 2008), or something similar to interests or preferences (Pouteau 2014, 9), or enjoy a state of *well-being* when cared for or left alone to flourish (Kallhoff 2014, 685–700) and can thus be *harmed* or *disrespected* in some philosophically interesting and non-trivial way, is definitely worth exploring.

The task for this latter sort of plant ethicists is to argue, based on the deployment of a key moral concept like autonomy or dignity and the mobilization of the classical justificatory apparatuses that correlate to these concepts (e.g.

Kantian ethics, virtue ethics, intersubjective ethics), that the Sugar Maple or the jade plant are as worthy of our moral duty, care or consideration as any animal analogue. This position is supported by the everyday gestures of the careful gardener as well as by political institutions (the moral standing of plants is enshrined in and protected by the national constitutions of Switzerland and Ecuador). Yet, it is still common to hear ethical theorists who have been working on the question of plant ethics admit that "granting moral standing to plants remains a difficult endeavor at this stage" (Pouteau 2012, 154).

Another angle altogether?

Any ethics whose origins lie in the humanist realm, or which works with concepts developed for and applied first and foremost to the human subject, may possibly be made to work (at least to some extent) for animals, because humans are animals, but it is not going to work for plants because plants are *really* not animals. Attempting to get moral insights on plants by reasoning in reference to the animal exemplar – especially now that much is made of animal *minds* and *cognition* – is ultimately an obstacle on the way to a wider ethical imaginary and surely to the de-centering of the human subject as *the* moral subject and object. It is an obstacle in two senses, one more obvious than the other.

The first is that such an ethical imaginary continuously ratifies the human subject as the exemplar: even when sincerely searching for (empirically sound) vegetal variants on *dignity* or *interests* or *inter-subjective self-hood*, the sense of these very terms is modeled on what these terms mean for us and to us as minded agents (e.g. conscious, cognitive, intentional, belief-forming, and desiring) (Houle 2010, 12–13).

A cursory glance at the developments in animal ethics of the last 30 years confirms this claim. The degree to which non-human animals look or act or sound like humans is directly proportional to the confidence with which ethicists (drawing from comparative cognitive science, social cognition, animal cognition, behavioral studies, and linguistics) can state (and get away with the statement) that these perceived similarities really correspond to features of the animals under consideration, as opposed to being mere projections (Andrews 2012).[2] Extending that angle of approach from the animal analogue to the plant kingdom seems even less promising, for "considering morphogenic forces embodied in biophysical forms and organic performances of plants, it can be shown that plants exhibit completely original features that make them incommensurable with animals" (Pouteau 2012, 154). This insight forces the development of an ethics *for* and *about* plants which in no way derives from, or depends upon, similarity of function or value or morphology, to animals: an ethics of difference.

The second obstacle is an even deeper impasse at the level of our very systems of moral thought. As I have argued elsewhere (Houle 2018), my efforts at thinking plant ethics begin with the ontological hypothesis that the concepts with which we think are formed, or have form. Concepts are the body of thought, and systems of concepts are thought's body. Like bodies, concepts are inherited, undergo morphogenesis, and are subject to evolutionary pressures; and so, over time and use, give rise to a stable, patterned being-of-thought: what thought is and how thought functions. This is another dimension of what Foucault called "a regime of truth" (Rabinow 1984, 55).

Because we humans are thinking beings with animal bodies, enfleshed and in the world, with a distinct biophysical form and organic performance, our thinking parts – our concepts – are also constituted by the patterns and forms and habits of performance that are characteristic of us *qua* embodied beings. A Spinozist ontological parallelism permits us to imagine that our thoughts have, or are, also bodies. A thought-body. That means that moral concepts too – the concepts of *dignity*, *integrity*, *harm*, *rights*, *care*, *virtue*, etc. – have, exhibit and function in animalesque ways *as forms and styles of thought*. What this means is that we can never even get to the vicinities of a moral concern for plants if we set out equipped with ideas that can only ever orient themselves, and move, and reproduce, and react and behave as animal-bodies do. It is possible that some deeply ingrained mechanisms in our conceptual apparatus and thinking behavior prevent us, as Levinas might say, from ever being actually confronted by the ostensible object of ethical concern: "the worldhood of other existences than that of human beings" (Marder 2012, 264). This would confirm Pouteau's reflection that "the animal seems to stand more as a theoretical block than a useful reference for moral consideration of plants" (Pouteau 2012, 155), and also deepen it by adding that "the animal form of thinking" is the very *form* by which our moral *consideration in general* happens.

My own philosophical interest in plants ethics finds its inspiration in this very impasse. For although I care very much about plants, such as the jade plant on my desk and the lovely *Acer saccarum* across the street, I am drawn to the work of thinking-about-plants-ethically less as an ethicist than as a philosopher more generally. What interests me is what this line of inquiry, diverted from its more customary angles of approach, might accomplish not for plants, but for *thinking itself*. The second sense that I have identified in which "animal ethics" is an obstacle to post-humanist ethicality may inhibit the expression of a fuller range of our capacities as thinking beings. When we automatically deploy, as our ethical thoughts, our animalesque repertoire of concepts, we miss out on an opportunity by which thought itself might mutate upon encounter (with plants, in our case) and might *become*.

A condition for the creation of concepts is the appearance of a genuinely new problem. The genuine heterogeneity of plant and animal modes of existence indicates that plant ethics really is a new problem for ethics and ethical thinking. The radical nature of this challenge has been largely underappreciated. The inadequacy of animal-based ethics for discovering and grounding our obligation to vegetal life is indeed a unique moral problem in need of its own concepts or, at the very least, it is reason for a significant revision of our traditional moral concepts. But in addition to this, the radical challenge of vegetal existence *for philosophy* is that it offers an occasion for recasting styles of reasoning that are endemic to traditional Western ethics itself. Such recasting may be necessary:

arguably, our moral concepts and styles of reasoning have proven unable to negotiate not just the challenges of plant ethics but also many other real challenges to the quality of human existence and even to human existence itself: poverty, hunger, war, terrorism, colonization, racism, biodiversity loss, climate change, among others.

A plant ethics that diverts from the animalesque path, and meets and works through the *worldhood* of plants, holds the promise of a *becoming of thought* or, at the very least, of testing the healthiness of our concept-bodies (Houle 2015, 397–414). In addition, insofar as the becoming of thought or a healthier concept-body enables us to respond more subtly and vitally to the full array of *what it is* that presents itself to us as questions of value and meaning, wherever they arise, then such becoming of thought is in itself an ethical event as well. Pouteau notes the same radical potential:

> Finally, history seems to be making a feed-back loop: plant domestication was the foundational act of agriculture and moral consideration of plants is now making this foundational act a matter of conscious, cognitive comprehensiveness [. . .] it is proposed that, as a mirror-image, the moral consideration of plants may involve a mental revolution potentially as paramount as the "Neolithic revolution."
>
> (Pouteau 2012, 158)[3]

Animal bodies, conceptual pathologies

As I have argued in *Complexity, Responsibility and Abortion: Toward a New Image of Ethical Thinking* (Houle 2014), "contemporary ethics" is absolutely dominated – in pedagogy, in scholarship and in practical training such as one finds in professional societies and research ethics boards – by a ubiquitous, deeply oppositional style: Which of these *two and only two* positions – for example "Pro-life" or "Pro-choice" – should I take? How can I defeat the other position? Which framework offers knockdown arguments? Is euthanasia for the terminally ill *right or wrong*? Should I attempt to save the drowning child or should I not? Should I pull the trolley car lever or should I not? Are harms reduced or increased by making the research survey anonymous? And so on.

All of the concepts we currently use to think *about* questions of value including abortion, euthanasia, poverty, or the livability of the planet – concepts like integrity, dignity, freedom, compassion, responsibility, harm, justice, and fairness – are marked by, express, and hence operationalize some variant of one or several of the same patterns expressed and operationalized in our basic zoomorphology: our animal body-plans or their morphological (developmental) patterns. These are *verticality* ("higher vs. lower"; "stronger vs. weaker"); *strict bilaterality* ("right vs. left"; "right or wrong"); *dorsal-ventrality* ("forward-thinkers vs. primitivists"; "front-runners vs. laggards and freeloaders"); *linearity* ("now vs. later"; "cause and effect"); and *centripetal-centrifugality* ("inner circle vs. marginal"; "origins vs. outcomes"; "essential self vs. dissociated self"; "subject" at remove from its "object" out there). Since these are dominant patterns that evolve and stabilize across time, our styles of ethical reasoning and operations of ethical thought tend to become fixed by these same patterns, which are thus their "*how*" (Foucault 1983). For instance, our first question about the blame and culpability for an unfortunate event is what caused it. The questions "*why?*" and "*what?*" are usually the first questions we ask about the meaning of events (see Deleuze 2006, 76). The very intelligibility of these questions, in turn, rests on a prior formal operation by which the past is conceptualized as standing in linear relation to the future. Our next question is almost always: *who?* Of all the forces, factors, and beings related in some way to the unfortunate event, whom among them should be blamed? Again the very intelligibility of the question of blame rests upon three prior formal operations, which are built in the very concept of blame: verticality (according to which one more blameworthy rises above the many); centripetal force (which produces a unified moral object – a *what* – out of a dispersion of effects (Foucault 1972)); and centrifugal force (an innermost moral will, a motive, a genetic defect, desire, or vice exists in the moral subject that is at the centre of the action).

With respect to this last point, Foucault called the mechanism at work "a doubling" – whereby a unified type of isolated being is *produced* through the machinations of the forms of power that our modern discourses, such as law and medicine, operate with:

> [. . .] the offense is doubled with a whole series of other things that are not the offense, but a series of forms of conduct, of ways of being [. . .] that are then the cause, origin, motivation and starting point of the offense.
>
> (Foucault 2003, 15)

What Foucault spots here is a nearly invisible "*how*" whereby the *idea of an interiority of truth* (e.g. personality, type, character, and psychology) is proposed and ratified by the very structure of the intelligibility of discourse, and *then* becomes visible in concrete practices such as "expert testimony" in court cases and in explicit concepts such as "mania" and "degeneracy" which then appear as *the reason* why the individual did what he did: the cause was inside him all along. Law and medicine make a living human into a unified thing with a depth, an infinite and powerful inside: a *who* that is the *kind of thing* that does what he did and (thus) is responsible for it:

> What do these expert testimonies reveal? The illness? Not at all. Responsibility? Not at all. Freedom? Not at all. There are always the same images, the same actions, the same attitudes [. . .]. "He played with wooden weapons." "He cut the heads off cabbages." "He was a trial to his parents." "He played truant from school." "He didn't learn his lessons." "He was lazy." AND THEN: "I conclude from this that he was responsible."
>
> (Foucault 2003, 15)

Now, the proper diagnosis is not that these conceptual body-plans are themselves bad. Following a line of enquiry that goes from Spinoza through Foucault and Deleuze, our real interest lies in what these forms can accomplish and in what they fail to enable. Our interest is always in what they can and cannot do, and what thought-actions they effect upon other thought-actions.

Since we are animalesque in body, including in the bodily capacities of our concepts and styles of reasoning, it is no surprise that we are pathologically wrapped in hierarchical, dualistic, bilateral, oppositional, judgmental forms of ethics. The animalesque lineage of our ethics is similar to an extremely well-tended cognitive topiary. Alternative forms, styles, or capacities of ethical responsiveness are atrophied; and for this reason we often do not know how to open ourselves to, inhabit, and respond to different ethical modalities, of which there are many. Some examples of such different modalities are aporetic post-structuralist ethics, the ethics of sexual difference of Luce Irigaray, or the beautiful ethics of infinite responsibility developed by Levinas:

> [. . .] a responsibility that goes beyond what I may or may not have done to the Other or whatever acts I may or may not have committed [. . .]. A guiltless responsibility, whereby I am nonetheless open to an accusation of which no alibi, spatial or temporal, could clear me.
>
> (Levinas 1979, 83)

Our conceptual atrophy is especially egregious when the phenomenon calling for ethical response on our side is simply unlike the situations to which our usual responses are ontologically adequate or epistemically apt. There are phenomena that simply do not bear the ontological features of being rankable, serializeable or bifurcatable. In these cases, unreflectively deploying stock normative concepts (with their latent cutting operations such as guilty vs. innocent; right vs. wrong) is intellectually inappropriate and possibly morally disastrous – as doing so prevents us from "taking account." Timothy Morton has made this argument regarding climate change (Morton 2013). Besson says something similar about organic agriculture (Besson 2009, 329–62), and Emily Yates-Doerr about global hunger and food security (Yates-Doerr 2015, 229–46). I have approached the phenomenon of unwanted pregnancy in this way (Houle 2014). Regradless of the particular phenomenon that is at stake "this [binary] way of thinking about and practically approaching these sorts of issues [. . .]" (Biro 2015, 36) blocks the ethical imagination from encountering the ethical that is at issue –, that is actually seeing or sensing what we are faced with. This is true also of lower animals, not just plants:

> People often ask whether "lower" nonhuman animals such as fish or dogs perform sophisticated patterns of behavior that are usually associated with "higher" nonhuman primates [. . .]. In my view, these are misguided questions [. . .] because animals have to be able to do what they need to do

> in order to live in their own worlds [. . .]. I want to reemphasize that the use of the words "higher" and "lower" and activities such as line drawing to place different groups of animals "above" and "below" others are extremely misleading and fail to take into account the lives and worlds of the animals themselves.
>
> (Bekoff 2008, 152)

Synthesis

Several threads of our inquiry can now be brought together. A first challenge for plant ethics is how to disrupt or divert the overly tenacious animalesque capacities of our moral consciousness, our conceptual apparatus and styles of ethical thought. A second challenge is to select alternative ways to look at the very different "lives and worlds" of the very different beings that plants are. A third challenge lies in individuating these ostensibly distinctive qualities of vegetal existence that make "plants radically and unconditionally different from animals" (Pouteau 2012, 155).

My suggestion is that it is precisely to the distinct "worldhood" of plants that we should expose ourselves in order to *become more adequately ethical in and to the world, in its incredible diversity.*

My synthetic method is to bring what the natural sciences (e.g. botany, biophysics, developmental biology, and ecology) have to say about the peculiar qualities of vegetal existence into direct, productive proximity with philosophy. This is one way to expose our concepts (especially our normative concepts) and ways of thinking (especially in our valuing) to these other lives and worlds. This "hermeneutic new realism," in the tradition of Goethe and Von Uexküll, is committed to finding the significance of plants for ethical thought without helping itself to external referents such as God or humans: it passes immediately to the natural forms of plants, readily on display in observed plant life itself. However, my method is not characterized by a reverence for the natural sciences and much less is it a requiem for ethical inquiry. It rather ought to be seen as equally an effort to "raise science beyond its currently limited scope [emphasis added]" (Pouteau 2014, 18–19). This approach reveals the many non-superficial and philosophically trenchant ways in which sessile vegetal existence differs from animal existence.[4] I offer an example of this approach in the next section, which focuses on but one aspect of plant bodiliness: the fact that plant bodies are not bifurcated in the same manner or to the same degree as animal bodies: plants are not divided "internally" or "externally," as animal bodies are. Our acknowledgement of this – I mean, really deeply understanding this one incredible fact – already sets us on a more promising path toward "plant ethics," insofar as such information itself becomes "internalized" not merely as an fact (the content of thought) but as an in-forming element of a change in the form that our moral concepts take as they engage with plants, and a radical rearrangement of the formal relations among these various moral concepts.

Plant bodies, unsplit surfaces

Plant architecture is currently the "hottest area of plant developmental biology" (Wang and Li 2008, 253–79). All plant and animal bodies develop from a sphere: higher plants from a pollinated seed; humans from a fertilized egg. A sphere is a maximally symmetrical object: any plane that passes through its midpoint would still preserve its symmetry. A sphere also faces all directions at once. Fertilization is the beginning of a complex event that involves a change from a sphere-like object into an object-body with directions: an *up* versus a *down*, a *right* versus a *left*, and an *in* versus an *out*.

In a mature plant embryo there are distinct zones comprised of a first-order cell polarity labeled SAM (the "shoot apical meristem") and RAM (the "root apical meristem"). Throughout the life of the plant, certain cells reach up this apical-basal axis, while others only ever reach down. The up and down of plants is maintained or regulated by a stiff plant cell wall. The cellular division along this polarity occurs by elongation, establishing a permanent vertical axis in the plant.

If the plant protoplasts are isolated from a plant cell, the embryo will regress to "a sphere of tissue" sensitive to environmental cues (Hudson 2000, 356). One speculation is that the external cell wall of all plant cells is a descendent of the original external wall of the zygote, so that "all plant cells inherit a determinant of outside identity" (Hudson 2000, 355).

How is this like or unlike the spatialization of the animal embryo and, later, body? At first glance, it looks the same: humans, whether in the early embryonic stage or later in life, also enjoy an up-down axis. Like plants, we elongate along that axis, and maintain verticality all our lives. However, since animal cells lack the same sort of stiff cell wall that plant cells have, it must be the case that the up- and downness of our bodies is established and maintained by some other forces. It is posited that our verticality comes from a rigid internal element – our spinal cord – rather than from the way that our most outer cells – the cells of our skin – are perpetually receptive to environmental cues of directionality. A permanent, immanent total-self-shaping relation to proximate otherness is not found in us but is a special feature of vegetal ontology.[5]

The second axis of directional development within the embryo, whether animal or vegetal, is radial: our roundedness and our tubularity. In the plant, distance builds between the center and the periphery, a pull outwards whose overall morphological result is a concentric thickening (Laux, Wurschum, and Breuninger 2004, 190–202). However, in order for a horizontal elongation to occur there must be a greater "radial than tangential velocity of growth" (Thompson 1961, 297) toward the outside, and a spreading of this thickening along an emerging arc so that the periphery becomes continuously rounded. The horizontal and the vertical axes of growth operate independently in plants, and yet they must work in an integrated and coordinated manner. In other words, two entirely separate and distinct planes of dynamic growth must sync to build a wholly integrated 3D plant body out of rounding and elongating actions.

However – and this is a crucial point for us – this coordination of entirely different symmetries does not lead the plant to have the same disjunctive inward- and outerness that we can observe in a developing animal body (Hudson 2000, 349–70). Animals undergo an additional, major change in shape early on in embryogenesis, which marks a disjunction between the inward and the outer. This change, called gastrulation (Pouteau 2014, 19), consists in an invagination of the embryo and the creation of an actual inner space: an empty tube. This inner space is, in fact, an internalization of what was initially facing outwards. The inside of animal bodies – mouth cavity, throat, esophagus, stomach, intestines, and anus – is the outside world turned inwards. This mechanism of gastrulation "gives rise to [. . .] an actual dualistic state of being" (Pouteau 2014, 19). Plants do not have this dualistic state. As Hegel put it in *The Philosophy of Nature* (1830, 1970, 307), plants have "a negative, simply unity . . . an inner indifference". Plants live as "unsplit beings (having neither an inside nor outside); "non-topos" in an undivided, unlimited, non-centered state of being" (Pouteau 2014, 20). Plantness faces only outwards (Pouteau 2012, 156).

The ethical implications of this hermeneutic realism are profound. First, one of the greatest obstacles to an adequate environmental ethic is human arrogance: an ego that thinks it is both better than and separate from everything else. Second, how could such an ego-structure (and its concomitant arrogance) ever be upended, replaced by a radically different way of being in the world and thinking in, and as that world? A proposition I have been exploring is that learning and understanding the radical uniqueness of plant morphogenesis is the exercise of practicing a different way of being a relation of self and world.

Conclusions: Opening the how

Could we – for example our concepts, our thinking, our norms, and our bodily sensibilities – learn to face only outwards and in all directions at once, like plants? Could we train ourselves to respond differently in ethically charged situations if we augmented our *habituated* propensity for binary reactivity with efforts from entirely other forms and styles of response? Reacting is not the same as responding. We are not *responding* – mentally, conceptually, spiritually or ethically – to complex phenomena, such as an unwanted pregnancy or other "lives and worlds," like those of plants, if anytime we are confronted with them we automatically resume "our positions" and repeat the judgments that follow from these positions.

From my work on the incendiary ethical issue of abortion, I can attest that the way the issue is confronted is, on the one hand, marked by strict binary oppositions and, on the other hand, affectively charged by extreme reactivity and dispiritedness. The uncompromising and dark qualitative tone, or style, with which this deeply meaningful human phenomenon is often treated might not be due to fetal viability, or patriarchy, cultural or financial resources, effective birth control, women's reproductive autonomy, the far right, colonialism, charismatic leaders, or technological developments in neonatology. It might rather be an effect of very rigid, habitual forms of thinking and valuing which prevent us from sensing the ethical vitality of the issue, engaging honestly with that vitality, and drawing creatively upon the special intensities that traverse it.

What if our sense of self, both individual and collective, could live at the heart of us, in our depths and, at the same time, flourish in the immense variation that dwells, permanently and continuously, on our living surfaces? Could we not grow and thrive, and live well as an "indeterminate ontology, [without] clear spatial and temporal boundaries [. . .] never [. . .] conceived of as complete?" (Pouteau 2012, 156).

Notes

1. "[. . .] among non-humans, and separate from non-humans, there is an immense multiplicity of other living things that cannot in any way be homogenized, except by means of violence and willful ignorance" (p. 48). Derrida is speaking of the multiplicity of the non-human animal, but his edict applies equally to every other non-human living thing.
2. In a talk at the Campbell Centre for Animal Welfare at the University of Guelph (October 2015), while discussing the current discursive field in animal comparative psychology, Andrews underlined that the burden of proof falls squarely on those who claim similarities to give scientifically rigorous evidence that what they claim to be seeing or recognizing in an animal experiment ("beliefs", "understanding", "intentionality", etc.) is indeed that and not a pure mechanism or a projection. An inability to offer philosophically and methodologically rigorous defences and explanations of observed phenomena threatens these claimants with the epistemic downgrading to mere "folk psychologists". Andrews adds that even while describing those animals most like us (the Great Apes), it is intellectually respectable to deny that what we are seeing is "friendship" (Jonas-Suilk) or "grief" (Eckmann). Be that as it may, my point here is that such epistemic mechanisms of projection may have the effect not only of keeping other animals (let alone plants) in a state of deferred and "questionable" subjectivity for the purposes of ethics, but also reinforce a way of thinking *about ethics* that assumes our human lives and selves to be the meaning-making epicentre of moral phenomena like friendship and grieving.
3. In this paper, Pouteau asks what the moral concept of "integrity" might come to mean once it is actually exposed to the complex facticity of vegetal life. She reasons that it cannot and does not retain a resemblance to the concept as we now apply it to an animal body or animal life.
4. I have used this approach with plant signaling (see Houle 2012, 183–94).
5. Epigenetics (the study of the environment conditions under which genes are turned on or off) is corroborating this explanatory shift from internal causes to external causes; or to a combination thereof (see Cubas et al. 1999, 157–61).

References

Andrews, K. 2012. *Do apes read minds? Toward a new folk psychology*. Cambridge, MA: MIT Press.

Bekoff, M. 2008. Animal rights and the great ape animal project. Pp. 151–6 in *The animal ethics reader*, edited by S. J. Armstrong and R. G. Botzler. New York, NY: Routledge.

Besson, Y. 2009. Une Histoire D'exigences: Philosophie et Agrobiologie, L'actualité de la Pensée des Fondateurs De L'agriculture Biologique pour son Développement Contemporain. *Innovations Agronomique* 4: 329–62.

Biro, A. 2015. The good life in the greenhouse? Autonomy, democracy, and citizenship in the anthropocene. *Telos* 2015(172): 15–37.

Bowman, J. L. and Floyd, S. K. 2008. Patterning and polarity in seed plant shoots. *Annual Review of Plant Biology* 59: 67–88.

Cubas, P., C. Vincent and Coen, E. 1999. An epigenetic mutation responsible for natural variation in floral symmetry. *Nature* 401: 157–61.

Deleuze, G. 1988. *Spinoza: Practical philosophy*. San Francisco, CA: City Lights Books.

Deleuze, G. 2006. *Nietzsche and philosophy*. New York, NY: Columbia University Press.

Derrida, J. 2008. *The animal that therefore I am*. New York, NY: Fordham University Press.

Foucault, M. 1970. *The order of things*. New York: Pantheon.

Foucault, M. 1972. *The archeology of knowledge*. New York, NY: Pantheon.

Foucault, M. 1983. The subject and power. Pp. 208–26 in *Michel Foucault: Beyond Structuralism and Hermeneutics*, edited by H. L. Dreyfuss and P. Rabinow. Chicago, IL: University of Chicago Press.

Foucault, M. 2003. *Abnormal: Lectures at the Collège de France 1974–1975*. New York, NY: Picador.

Hall, M. 2009. Plant autonomy and human-plant ethics. *Environmental Ethics* 31, no. 2: 169–81.

Hegel, G. W. F. 2004. *Philosophy of nature: Encyclopedia of the philosophical sciences, Part II*. Oxford, UK: Oxford University Press.

Houle, K. 2010. Infinite, indifferent kinship. *C Magazine* Autumn 2010(107): 12–23.

Houle, K. 2012. Devenir plante. *Chimères* 2011(76): 183–94.

Houle, K. 2014. *Complexity, responsibility, abortion: Toward a new image of ethical thinking*. Lanham, MD: Lexington.

Houle, K. 2015. Toward a healthy concept of health. Pp. 397–414 in *One health: The theory and practice of integrated health approaches*, edited by J. Zinsstag, E. Schelling, D. Waltner-Toews, M. Whittaker and M. Tanner. Oxfordshire: CABI International.

Houle, K. 2018.Symmetry and asymmetry in conceptual and morphological formations: The difference plant body growth can make to human thought. In *Deleuzian posthumanisms*, edited by C. Daigle and T. McDonald. Bloomington, IN: Indiana University Press and SUNY.

Hudson, A. 2000. Development of symmetry in plants. *Annual Review of Plant Physiology and Plant Molecular Biology* 51, no. 1: 349–70.

Kallhoff, A. 2014. Plants in ethics: Why flourishing deserves moral respect. *Environmental Values* 23, no. 6: 685–700.

Laux, T., T. Wurschum and H. Breuninger. 2004. Genetic regulation of embryonic pattern formation. *Plant Cell* 16: S190–202.

Levinas, E. 1979. *Totality and infinity: An essay on exteriority*. The Hague: Martinus Nijhoff Publishers.

Marder, M. 2012. The life of plants and the limits of empathy. *Dialogue* 51, no. 2: 259–73.

Morton, T. 2013. *Hyperobjects: Philosophy and ecology after the end of the world*. Minneapolis, MN: University of Minnesota Press.

Pouteau, S. 2012. Providing grounds for agricultural ethics: The wider philosophical significance of plant life integrity. Pp. 154–9 in *Climate change and sustainable development: Ethical perspectives on land use and food production*, edited by T. Potthast and S. Meisch. Wageningen: Wageningen Academic Publishers.

Pouteau, S. 2014. Beyond "Second Animals": Making sense of plant ethics. *Journal of Agricultural and Environmental Ethics* 27, no. 1: 1–25.

Rabinow, P. 1984. *"Truth and Power" the Foucault reader*. New York, NY: Pantheon.

Smith, D. N. 2012. Surfaces, painterly illusion, metaphysical depth. *A Journal of Modern Critical Theory* 35, no.3 (November): 389–406.

Thompson, D. W. 1961. *On growth and form*. Cambridge, MA: Cambridge University Press.

Wang, Y., and J. Li. 2008. Molecular basis of plant architecture. *Annual Review of Plant Biology* 59: 253–79.

Yates-Doerr, E. 2015. The world in a box? Food security, edible insects, and "One World, One Health" intervals of confidence: Uncertain accounts of global hunger. *Biosocieties* 10, no. 2: 229–46.

7 Plants as open beings

From aesthetics to plant–human ethics

Sylvie Pouteau

Why is plant ethics not yet an established field of research, in the same way that animal ethics is? Given the enormous amount of human activities that rely on direct or indirect interaction with plants, the issue cannot be dismissed as merely futile. Furthermore, because we all take part in a shared, interconnected web of intended and unintended effects unfolding on an evolutionary scale, living entities that thrive and sustain their own ways, such as plants, necessarily have their share in our own human ups and downs. Even if one believes that plants are an extremely low category of slaves meant to satisfy human needs, the moral implications of this slavery ought not to be passed in silence. In my view, major difficulties for plant ethics lie in the way that we conceive of plants as plants, and more generally in the way that we frame ethical issues with reference to a modern mindset. To perceive and address specific loci of value that lie beyond human beings and animals, we need to break with a number of typically modern conceptual gridlocks, and accommodate more recent, postmodern understandings of complexity and entanglement.

While discussing the postmodern deconstruction of the Cartesian subject, John Baird Callicot stated: "Step by step, the subjectivity of the subject is eroded until, with plants, we reach self-valuing nonsubjects" (Callicot 1995,74) In spite of its negative connotation, the term "nonsubject" is quite interesting because it contains a formal contradiction: it is a subject *and* it is not a subject. In other words, we hardly know anything about this type of subject – we do not even have a word for it. Put another way, non-human(-like) entities like plants cannot be addressed on the same epistemic grounds as human beings and their animal kin: they require their own specific means of ethical investigation.

In the last decades the plant has become emblematic in postmodern thought: a truly postmodern subject. To a large extent, this view has been propagated by the influential notion of *rhizome* introduced by Gilles Deleuze and Félix Guattari (Deleuze and Guattari 1980). As a rhizome, the plant nonsubject is decentralized, net-like, connectedly robust, and it is a multiplicity at "n – 1". In my view, this multiplicity at n – 1 is the most significant feature of this definition. While sequential addition of parts at "n + 1" may suggest a progression towards some more completed state, subtractive multiplication at n – 1 emphasizes the fact that plants are open beings never completed and (potentially) endlessly coming into existence. To account for this open "always $n - 1$" character, I believe that a new type of realism needs to be explored based on an aesthetic-pragmatist approach (Pouteau 2012).

In the first part of this chapter I introduce a new realist definition of the plant kingdom, considering the way in which the plant's open form behaves. In the second part I examine the implications of designing the open form through the processes that underlie plant domestication and the creation of plant–human forms. In the last section, I discuss how caring for open beings can make sense, and how it can be promoted by exploring aesthetic features of non-centered plant life.

Distinctive ontological habilitation of plants: A pragmatist position

Kingdoms, theory and experience

To define the distinctive character of plants, I will first state that open life is a kingdom, though not in a taxonomic sense, and I will adopt a pragmatist position to explain why. In the taxonomic sciences, a kingdom describes the second highest rank of classification of the biotic community, below a domain. The number and frontiers of biotic kingdoms have changed several times through the history of taxonomy. Recently, for example, fungi have been moved out of the plant kingdom and grouped in a separate kingdom (Margulis and Schwartz 1982). Interestingly, one frontier that has never really been challenged so far is that between animals and plants.

A first reason is that the distinction between plants and animals cannot be addressed because the animal *is* the form by which plants happen to be a subject of science. In fact, the whole field of plant sciences emerged in the 18th century because it was discovered that plants are composed of the same organic matter as animals and therefore that they can be compared to animals (Delaporte 1979). At the same time, these scientific advances also prompted new moral concern for plants. The moral issue of plants has always been dominated by the theoretical model of the animal: "the animal is the *form* by which our moral *consideration* happens" (Houle, in this volume). During the 20th century this theoretical model has been extended to the whole biotic community and condensed in the term "biodiversity", with the notion of *bios* ("proceeding life") by and large haunted and cannibalized by its animal counterpart, the notion of *zoè* ("motion of life"). To a large degree, (plant) biology is still at that stage: a kind of zoology, or "zoèlogie".

I have argued elsewhere that theoretical arguments can only lead us to seeing the plant as a "second animal", or as a "non second animal", because these arguments are cognitively indexed to the animal archetype (Pouteau 2014). The ironical comments that accompanied public discussions on the dignity of plants in Switzerland (ECNH 2008) have been quite instructive in this regard. The irony mostly fed on absurd anthropomorphisms and animalizations of plants.

The sarcastic message was that common sense dictates that a plant is simply not an animal or a human – and the unexpressed premise was that moral concerns could only be raised if plants could in some relevant way be equated with animals and humans. This was all presented as obvious. These apparently obvious ways of reasoning are paradoxically illuminating, because they reveal how much of a blind spot plant life currently is for common sense. In fact, these ways of reasoning tell us nothing about what a plant is, except that it is obviously not something else.

A second reason why the distinction between plants and animals is typically not addressed is that the notion of kingdom itself has been given little attention so far. In fact, the scientific community has been mainly concerned with the notion of species (still entangled in the Platonic, essentialist tradition), and with working out new definitions of species on evolutionary grounds (Mayr 1957; Ruse 1995). One interpretation is that the notion of kingdom is thought to be simply a surrogate of the notion of species at a higher level. A more radical interpretation is that the scientific emphasis on the notion of species fails to acknowledge another strongly established classification of the natural world in four kingdoms (mineral, plant, animal and human) that has been influential in Western tradition at least since medieval times. Thus, the distinction between plants and animals is pivotal because it links the two kingdom classifications together.

I think that what is at stake when considering the distinction between animals and plants is the old divide between nominalism and realism on the issue of universals (MacLeod and Rubenstein 2006). From a scientific, nominalist standpoint (the one used in taxonomy), the difference between plants and animals is only one of degree. But from an ordinary, realist standpoint, plants and animals are natural kinds and belong to different orders of reality. Here, I propose that we define a kingdom in a new realist sense in order to reconcile ordinary experience with current scientific knowledge (in the following sections I will refer to kingdoms only in that sense). I believe that a pragmatist position on life can lend support to this definition because pragmatism bears the notion that experience, whether it is cognitive or sensory, is an interaction (Debaise 2007) and because action can be described on objective grounds in terms of performances or actual forces of physics (see, for example the notion of *Umwelt* coined by Jacob von Uexküll to describe animal worlds; von Uexküll 2010).

The French word for kingdom – *règne* – designates both a period of kingship as well as the act of wielding power. Rather than collections of objects that share similar features, kingdoms should be seen as expressions of actual powers in action. One cannot describe forms of natural entities merely as objects located in space. To account for their biophysical properties, one also needs to see them as processes, that is forms in action ruled by formative powers. Plant and animal forms are comprehensive architectures resulting from automatic auto-organization operated by fields of mechanical forces such as gravity and pressure. Biophysical analyses show that their formation involves distinctive laws, or rather, distinctive fields of hydrodynamic forces (Pouteau 2016b). This explains why plants and animals appear *obviously* different. In plants, the spiral arrangement of successive parts along a stem axis (called spiral phyllotaxis) is ruled by a field of axial forces

(Douady and Couder 1996). In contrast, the typical pattern of tetrapods, established after the blastula stage (an early stage in animal embryogenesis), is governed by a field of concentric forces (Fleury 2009).

The difference between fields of axial and concentric forces is radical, as it ensures that incommensurable types of beings are produced. An axial field of formative forces explains why the plant body remains open and continues (potentially) endlessly to come into being while declining at the same time. The fact that plants are sessile, green, photosynthetic entities is closely linked to this open character: plants do not need to move and feed because they do not have an "outside" (see below); and green may be described as the distinctive color of living beings that have no "inside", and thus no psychic life.[1]

In contrast, a field of concentric formative forces explains why very early during embryogenesis animals reach a closed, self-centered state bound to finiteness. Gastrulation is the ontogenetic event that establishes a boundary between an inside and an outside, an inner world and an outer world (Wolpert 1992). This divide leads to concomitant production of external parts and complex internal organization, and integration that support (potentially) endless possibilities of behavior.

Thus, because of the different fields of forces involved, plants and animals do not share the same dimensions of being in the world and no common measure can be established between them.

Incommensurability, a non-Euclidian space of flourishing

The open character of plants may be conceptualized with some approximation through notions such as growing, becoming (Houle 2011, 2012), or flourishing (Kallhoff 2014). Yet it is difficult to reach a fully comprehensive representation of plants as open beings, because a plant does not coincide with a completed object and cannot be assigned to a defined location in space (Pouteau 2016b). In contrast, closed, self-centered entities can be easily grasped with a typical Euclidian mindset. Our Euclidian concepts "are strongly marked by the animal form, or rather have an animal body" (Houle 2014, 160). In virtually all animals,[2] the division between an inside and an outside is bound to tridimensional (3D) patterning with three symmetry axes: vertical (top–bottom), horizontal (left–right), and sagittal (front–back). These symmetries act as oppositional pairs in a fully comprehensive system: for example left–right symmetry cannot be found independently from the other two symmetry axes.

As a comprehensive system, the 3D perspective relies on the notion that there is a unique focus point, a center that makes everything commensurable with any other thing. But apart from animals, most entities of the natural world, including plants, only have one or two symmetry axes (Hallé 1999). Thus, to conceive entities that only have one or two dimensions and to address their non-animal spatialization, one needs to seek another, non-Euclidian perspective.

To understand open beings, one has to figure out what it is to be a decentralized entity with no front or back, no left or right, but yet with a top and a bottom.

A truly a-dimensional "space" is actually beyond representation: without a one-dimensional (1D) polarity, conceptualization would be impossible. Land plants are the ultimate representatives of this minimal but foundational polarity: their top–bottom orientation is not a distinctive property of their own but rather a more general feature of the Earth–sun axis, explained by gravitropism and phototropism (Molas and Kiss 2009) Indeed, if tilted or put upside down, plants will change their orientation and strive to meet this Earth–sun axis again.

In addition to their 1D polarity, plants also exhibit massive two-dimensional (2D) organization and produce huge amounts of laminar plane surfaces, in particular leaf blades. Whilst 1D polarity is vertical and characterized by axial symmetry, the 2D organization expands horizontally with bilateral and proximo-distal symmetries (Hallé 1999). Because leafy layers only have two dimensions and no self-centered inside, they do not have true dorso-ventral symmetry. In fact, they enter into fractal compositions and occupy virtually no space while condensing enormous amounts of organic matter.

In addition to their 1D and 2D flourishing modes, plants also exhibit an inward motion exemplified by the spiral arrangement of appendages along the stem axis (Douady and Couder 1996).This inward motion can be described as the spiral principle. It explains the production of 3D-like structures by means of twisting, curling, folding and bending, and even the merging together of parts. In contrast to what happens with animals, this motion never leads to a proper invagination and remains unachieved and open. During reproduction, the spiral principle culminates in the principle of fructification characterized by the production of flowers and fruits. To some degree, flowers and fruits imitate 3D animal organization and may even exhibit dorso-ventral symmetry in addition to left-right and front-back symmetries (Cubas 2004).The most significant event is the formation of a pouch-shaped structure (called the ovary) by folding and merging together of laminar blades (called carpels) (Chiurugwi et al. 2007; Skinner, Hill, and Gasser 2004).This formation creates a pseudo-inner space in which local finiteness is achieved. But "inside" flowers and fruits there is actually no material for "inner" life, apart from seeds that await future development. Strictly speaking, seeds are located neither inside nor outside, since they never depart from the open like their mother plant. Flowers and fruits do not really have a centered inside and remain shadow images of true self-centered 3D forms.

How to define the center of an open being? Do plants have a proper center? "The quest for the secret source of plant subjectivity – that is, for a unified origin of its complex behavior – has become the new philosopher's stone in certain corners of botany" (Marder 2013). Centeredness is intrinsically linked to the notions of subjectivity and individuality. In my view, the individuality of "nonsubjects" can only be addressed by defining the standing of plant centers on non-Euclidian grounds. Because plants are open beings, their center must be located in the open, non-Euclidian space and not "inside" a putative 3D structure – be it a cell, a bud or a root-stock. Thus, the open character of plants calls for a re-assessment of the spatial premises that are used in life sciences, and for the re-positioning of issues linked to plant–human interaction. In the next sections, I shall explore what this may imply.

Domestication: Creating plant–human fields of forces

Energeia: The metaphysics of plant–human interactions

Physical bodies are forms in action enacted by *energeia*, that is, energy in action. In living beings, energy in action is supported by metabolism, that is the construction (anabolism) or destruction (catabolism) of chemical compounds. Though both animals and plants have anabolic and catabolic functions, their ontological relation to energy is radically different.

All energy flowing through the biosphere initially comes from photons provided by sunlight. Photosynthesis is virtually the only means by which this primal energy enters the biological world and this entry can only be achieved through open, anabolic beings – that is, plants (I will not discuss the case of algae and cyanobacteria here) (Gest 2002).Whether directly or indirectly, closed beings need to be catabolic parasites of open beings in order to gain access to the primal solar fuel stored in their body.

Plants do not need to move around to find their most vital resources. Their body "center(s)" is (are) already "around", whatever this "around" may represent in a non-Euclidian space. In contrast, animals (and human beings) lack everything they need to sustain their life, because their self-centeredness relies on the ontological division between an inside and an outside. Thus, to have a self-centered biological organization and to develop mobility and inner psychological life are consubstantial: these are two sides of the same coin. To have an inside means that one has to develop a more or less sophisticated "inner life" and the capacity to externalize this inner life through technical and symbolic outgrowths, that is to gain some degree of self-centered autonomy. Animals (and human beings) need to move around continuously and by all means in order to encounter the "outside", the non-self – that is fellow individuals and the rest of nature, and to recreate a sense of belonging to the world on both biological and psychological levels. Thus, motion is their most essential requirement and mode of expression. Here, the term motion includes local motility, both external and internal organic movements, *and* non-local motility relying on perception (emotions and sensitivity) and on representation (self experience and thinking).

The modification of the process that links the life of open and closed beings is more than a technical achievement and should be regarded as a major ontological shift in the course of evolution on the Earth. Plant domestication represents a true revolution in the history of mankind because it leads to the mastery of sun power and entails radical changes in Earth metabolism (Pouteau 2012). Commonly considered to be a lower category of human activity, food production is actually the catalyst of this global transformation. A major aspect of plant domestication is the enhancement of the fructification principle. It involves an increase in number, size and sweetness of fruits and seeds that do not fall but remain attached to mother plants until ripe. Moreover, it endows other plant parts with fruit-like properties: rounded shapes (obtained through folding, curling, swelling etc.), colors, and sweetness. Hence, roots, tubers, bulbs, stems, leafs, buds and inflorescences – for example beetroots, potatoes, onions, salads or cabbages – acquire the standing of "fruit". In many respects, the advent of agriculture can be described as the advent of a giant fructification process. In addition, vital human needs other than food, including heating, clothing and shelter, have been fulfilled by promoting the 1D- and 2D-flourishing of plants – as in the case of plants grown for fiber and timber production, or for animal feed.

The massive orientation of crop anabolism towards fructification means that availability of sugar compounds produced by photosynthesis is highly enhanced and that the energy input to human life can reach much higher levels. When modifying plant forms in action, human forms in action are in turn deeply transformed due to changes in diet and lifestyle. For all these reasons, to earn a living from crop culture is radically different from a lifestyle based solely on animal rearing. Fructification eventually expands to every aspect of human life at biological and techno-symbolic levels. Literally, crop breeding leads to new plant–human forms, that is forms in which plant life and human life have become tightly entangled and intertwined. Fields of formative forces of open and closed beings enter into new compositions so that the global form of the world is also modified.

The most striking expression of this intricacy is to be found in city life and its complex architectural, institutional and relational patterns. At this point, fructification finally brings the issue of energy to the forefront, in a way that may be equated to a new metaphysics of energy and sun power. A first reason is that energy is a metaphysical substance and does not coincide with a concrete stock of any substance but rather with a flow. Indeed, according to Aristotle there is a close relation between energy and motion (Brague 1990).[3] This epistemological difficulty has been well characterized for the notion of *qi* in traditional Chinese medicine (Fujimori, Yano and Hayashi 2013). A second reason is that the enormous complexity of techno-symbolic life obscures the actual significance of energy for closed and open beings, as well as the energetic implications of crop domestication. Thus, the link between the way we eat and our energetic needs, for heating and transport for instance, is barely addressed.

Designing the open, transforming the planet

The massive enhancement of fructification endows crop life with features of closed beings. Domestication may be seen as a kind of animalization process, or "becoming animal" (Deleuze and Guattari 1980). The expression "becoming animal" does not imply that crops have become a kind of animal or animal-like entities. It simply means that the animal-oriented flourishing mode has been favored over other plant modes of flourishing. Does this orientation mean that the ontological specificity of open life has been altered during the process of plant domestication? To explore further the significance of the animalization process for the domesticated standing of plants, I shall focus on two major features of this process.

First, crops now have a home and inhabit the human tridimensional space – for example lands and fields – rather than human beings inhabiting the plant open space – for example forests and grasslands. Having a home changes the innate relation between plants and land, as well as the standing of the land itself. Thereafter, the land may potentially become an unqualified, appropriated space – a space that can be virtually anywhere rather than a unique, ontogenetic place (Foucault 1967). The relation between plants and land may even come to rely exclusively on human practices.

Second, fruits and seeds are now left attached to mother plants and then kept in sealed places. Their motion is no longer ruled by un-fettered physical forces but depends (almost) entirely on human assistance and practices. This means that human activities, social rules and ethos are now integral parts of the open existence of plants. Meanwhile, being harvested changes the standing of fruits and seeds – which are no longer lines in a continuous flow of becoming but potential stocks of material that one can store over time.

With respect to these two features, the epistemic changes initiated in the eighteenth century represent a major turn. When plant scientists started to address plants in reference to a theoretical framework modelled on animal life, they also started to seek corresponding functions of finite, self-centered, catabolic beings: to eat, to breathe, to mate, to communicate, to move, and so on (Delaporte 1979). More or less at the same time, agriculture underwent a major revolution, with significant increases in crop production – so that the eighteenth century may correctly be described as the golden age of agriculture (Calame 2007; Claviler 2009). The new scientific conception of plants and technological advances in agriculture run parallel, probably feeding off each other. New attention was drawn to the organic relation between plant and soil and the standing of seeds by analyzing the two alternative modes of growth in living beings: size increase by means of nutrition and multiplication by means of reproduction.

Plant "nutrition"

The introduction of the notion of plant "nutrition" clearly relies on a metaphor, since plants do not have a gut and do not really ingest food. Nevertheless, the idea that roots suck or absorb soil nutrients makes sense, since practices that improve soil fertility result in increased production of plant biomass. The chemistry of nitrogen assimilation was only deciphered in the nineteenth century, but the agricultural revolution of the eighteenth century already relied to a large extent on practices that contribute to increasing nitrogen content in the soil: plowing, addition of animal manure, crop rotation (Calame 2007; Foster 2011). In the eighteenth century, however, these practices were no longer sufficient to maintain soil fertility while sustaining food production for growing urban populations. Soil depletion was so significant that organic fertilizers (bone powder, guano) had to be massively imported from other parts of the world. Karl Marx was one of the first to address the ecological issues of soil fertility and metabolic breaking as new major causes of concern (Foster 2011). Although the chemistry of nitrogen assimilation in plants was known by then, synthesis of nitrate was not yet at hand. Massive production of nitrogen fertilizers only became possible after the First World War, as a heritage of the development of the weapon industry.

The use of synthetic nitrate is not just one out of many agricultural advances, and should be regarded as a radical turn in agriculture. When "fed" with synthetic nitrate, plants are established as virtually soil-less beings. They get their "food" from factories, in the same way that sick people in hospitals are fed by drips. In some respect, the animal theoretical model has come true. Plants now have an outside that no longer belongs to them but is merely a substratum, a human artifact. As the organic link between the "outside" and plant life has been obscured, the soil itself is no more than a thing that can be appropriated and moved around. Crops can be fed to saturation so that nitrogen is no longer a limiting factor for growth.

Crops can also accumulate more sugars produced by photosynthesis and become "sugary fat". They thus become more appealing not only to human beings but also to a range of animals and microorganisms, thereafter called "pests". Enhanced nitrogen nutrition leads to a number of metabolic disorders in the larger system, which in turn need to be dealt with by applying a growing number of chemical drugs – in the form of pesticides and herbicides. Eventually, plant life is turned into permanent sickness and agriculture into an on-going war against numerous enemies.

Plant "reproduction"

Like the notion of nutrition, the notion of plant sexual "reproduction" relies on a metaphor, since plants do not penetrate or meet each others' insides (for the simple reason that they do not have an "inside"), and thus do not really copulate. In the animal kingdom, the egg is initially produced inside the mother's organism, even though its fertilization may occur outside. In contrast, the "egg" of higher plants is produced in a pseudo-inside that still belongs to their non-divided, open state of being.

During the eighteenth century, whether it was possible that plant sexual organs could be overtly exhibited in the open was still a controversial issue (Delaporte 1979). During the nineteenth and twentieth centuries, the interpretation that plants have a sexual mode of reproduction was finally favored based on the finding that plants and animals undergo similar cell and nuclear changes during reproduction. Nonetheless, a seed on a flower is more like an appendage than a fetus in a womb. The evidence supports "the idea that an ovule is a shoot-like structure (caulome) and that the outer integument may have evolved from a leaf or leaf-derived structure (phyllome)" (Skinner, Hill and Gasser 2004, 42).

To equate seed production with a different mode of growth ruled by the laws of a non-Euclidian space, rather than as a sexual mode of reproduction, may have far reaching implications for the regulation of seed trade. Indeed, without the assistance of human hands, domesticated seeds are usually unable to re-open new lines of becoming. In traditional societies, seeds exchange and gift have been necessary responses to the changes imposed on plant form during domestication. In contrast, when equated to a product of reproduction, the seed acquires the standing of a marketable object. This is reinforced by the necessity to breed crops that secure more stable production.

During the twentieth century, plant licenses and patents started to be implemented to reward the work of plant breeders and to regulate the seed market (Thomas and Bonneuil 2009). The free circulation of seeds was gradually reduced and is now tightly controlled. Seeds are locked-in and crops now have an "inside" in addition to their nutritional "outside" – but an inside that is unable to express any form of autonomous self-directedness. To a large degree, plants have now become animal-like entities.

Furthermore, what happens to crops eventually affects the whole planet – including all other forms of life, ecosystems, landscapes and human life. My conclusion is that concern for plants and concern for human beings and the rest of nature cannot be treated separately, as if they were independent issues. Adequate ethical approaches need to be developed in order to account for this entangled reality.

How to care for open beings

Non-centered ethics, plant–human ethics

At this stage, ethical concepts, words, and representations for capturing the specificity of open life are by and large still missing. Beings that are morally relevant should be self-centered and their moral standing should be established on criteria linked to this self-centeredness – for example reason, sensitivity, and agency. Since the Renaissance at least, self-centeredness has been established as the dominant form of spatial representation, so that everything should be seen through Euclidian lenses. This establishment permeates all domains of human life and operates through an "animal's body of thought" (Houle, in this volume). Even the notion of "system" may be reduced to a web of self-centered things located in Euclidian space.

An open being is quite different from a web of things: it is not a thing, it is a process. What the plant body exemplifies is the continuous unfolding, open-ended stream of life and evolution. Our difficulty in grasping the open character of plant life without using animal metaphors is more generally a difficulty in comprehending things that are not defined by an "immobile primary limit", that is, *topoi*.[4]

Henri Bergson pointed out that our concepts are formed in the same way as solid things, this is why "our thinking, in its purely logical form, is unable to figure out the true nature of life" (Bergson 1925,11). Although plants do not move in the proper sense, their topology changes continuously over time as if they were fluid things. To account for this peculiar, fluid type of *topos*, we need a more liquid form of thinking that is non-centered and peripheral.

In line with this open topology, ethical concern for plants should extend far beyond cost/benefit analyses, concerns for human health, and the like. It should be an integral part of a broader anthropological project regarding the ways in which plant–human societies operate at cultural, political and economic levels – for example land property rights, urbanism, landscape management, benefit sharing, fair trade, locality, work organization, conviviality, social participation, access to vital resources, conservation of nature, climate change.

To capture this entangled, cross-disciplinary complexity, I believe that the term plant–human ethics is more appropriate than plant ethics. John Charles Ryan also proposed the term human–plant studies to address the reciprocity that characterizes human–vegetal relationships and the need for developing interdisciplinary research frameworks to study it (Ryan 2012).

I submit that a major task for plant–human ethics will be to uncover ethical concern for plants in narratives conceived from within a Western topological mind-set. This is no easy task because in these narratives the emphasis is often not placed on plants themselves, but rather on the ways in which human life is affected by what we do to plants – for example narratives on soil fertility, pesticides, GMOs, diets based on high sugar content, seeds circulation, and so on. A diversity of protest movements concerning plants (some are listed below) seem to be good candidates for this enquiry because they had to elaborate narratives for their protest action – even if ethical concerns are often not explicitly expressed.

Thus, in the last two centuries, protest movements have been drawn to and explored local and circular approaches to agriculture and new ways of living together with crops, in order to address the issue of soil fertility and mitigate the metabolic alienation of the urban lifestyle. These movements include community-based agriculture movements such as the kolkhozes in the Soviet Union and the utopian garden city movement in England; as well as organic agriculture movements founded after the First World War, when chemical fertilizers started to be widely commercialized (Besson 2009; Calame 2007). More recently, farm seed networks based on participatory plant breeding practices have emerged in a number of countries, in opposition to the increasing power of the corporate seed sector, and with the purpose of maintaining free access to major crop gene-pools (Pimbert 2011). Numerous community-based initiatives linked to plant life have also started to flourish in urban areas – for example community gardens, the "incredible edible" network, the urban guerilla movement, and various forms of cooperative interaction between farmers and consumers in support of peri-urban agriculture (Baudry 2012).

Aesthetic experience of the non-Euclidian space

To make sense of ethical concern for plants, we need to address the non-Euclidian properties of their open body. A first step is to realize that our current Euclidian representation of the world is not univocal. A major factor for its stabilization has been the mono-focal geometric perspective that was invented in Italy during the *Quattrocento*. One thing that made the mono-focal perspective so popular was that it sized-up everything with a common measure overseen by a single principle: the vanishing point facing a potent monarch or subject (Latour 2015; Thuillier 1984).

This unique position has promoted major changes in science, architecture, urbanism, economy and politics – all changes that opened up the way to Modernity (Thuillier 1984). But this fecund "invention" can hardly be equated to a discovery of the way things are in reality (Arasse 2004; Thuillier 1984). In fact, its cultural stabilization could never be fully secured and in the nineteenth century mathematicians inaugurated a post-Euclidian era by inventing projective geometries, which ignore parallel lines and predict that space is curved, for example hyperbolic or spherical. In addition to having a different shape, these non-Euclidian spaces also exhibit distinctive properties, for example the capacity to act upon objects, which underpins the concept of relativity. At the turn of the twentieth century, these new geometries became known by a wider audience (Poincaré 1902). So far, however, they have not been integrated into standard educational curricula.

Non-Euclidian geometries could inspire possible ways of representing an open topology, but awareness is not merely a theoretical issue. Plants as open beings are non-Euclidian objects that require new forms of aesthetic appeciation as well. For instance, one needs to really perceive how plants grow as if they were pulled from a periphery that acts simultaneously on their multiple growing points located all over their surface. One should also explore how this is reflected in the actual ways in which we deal with plants in our most ordinary, sensory and kinesthetic experience; as well as in our body of knowledge, including the know-how gleaned in the gardening gesture.

To accommodate this need for aesthetic inquiry, I propose that plant–human ethics could greatly benefit from interdisciplinary interaction with artists and space designers. A first reason is that we need new ways of representing the obvious character of plant otherness. A major difficulty here is that the significance of open life is veiled by the fact that it is actually overtly exhibited in the open: "the secret is that there is no secret" (Marder 2013, 18). Johann Wolfgang von Goethe was probably the first to express this paradox by saying that nature is "mysterious in the open" (Hadot 2004, 160). According to Pierre Hadot, this statement inaugurates "a radical transformation of the notion of the secret of nature" (Hadot 2004, 263). It constitutes the cornerstone of Goethe's scientific work on plants, colors, and the weather: a new kind of realism in which aesthetics and art play an important role. To experience something that is obvious and hidden at the same time is a true philosophical event, an act of intuition that changes a primordial aesthetic experience into a thought experience. Artists could act as catalysts in this process.

A second reason why art and space-design are important to plant–human ethics is that spatial representation is a central subject of the dialectic between art and science. In the Renaissance, visual artists took advantage of the monofocal geometric perspective to explore new ways of representing what is actually beyond spatial representation – for example the mystery of incarnation (Arasse 2004). Subtle distortions in the geometric gradation have served to indicate that something there is incommensurable with human dimensions. Again, when non-Euclidian geometries became known in the twentieth century, artists rapidly saw in these spatial theories new opportunities for their aesthetic research, exemplified by the so-called "fourth dimension" paradigm (Henderson 2013). Again, artists could act as catalysts to explore non-Euclidian properties of plants.

Finally, a third reason why art and space-design are important to plant–human ethics is that in the last decades a growing number of artists have been exploring major issues linked to the use of natural environments and living entities in the fields of land art, environmental art, eco-art, and bio-art (Kagan 2010; Ramade 2007). A significant number of artistic propositions are closely related to plants and some of them address more specifically the issues of soil fertility, seeds circulation, and flourishing (Fel 2014). In this context, the scope of plant–human studies could be extended by research actions in which artists, scientists and philosophers can interact in the design of a method, language, or object and jointly elaborate new aesthetic notions and representations (Pouteau 2016a). Eventually, these approaches could serve as case studies to analyze the aesthetic significance of practices used in major areas of plant production, and the narratives of protest movements such as organic movements and farm seed networks. Finally, these research actions could provide new directions for developing more participatory approaches for the design of plant–human forms in the fields of agriculture and landscape management.

Further perspectives on open life

The open character of plants is not just one among many other factors to be considered in environmental ethics. Its investigation must link a wide array of epistemic, aesthetic and ethical issues. When considering the open character of plants, one is lead to explore a number of more general questions: (i) how do we recognize open life in nature, and how does this affect the way that we conceive of nature, the environment, ecosystems, and so on; (ii) do other beings fall into the category of open beings – e.g. fungi, algae, bacteria; (iii) how to re-appraise the mineral kingdom with respect to open and closed life; iv) how to reconsider open life when it is turned outside-in in animals and humans – e.g. our digestive tracts and the diversity of bacteria that they host; v) should human life be considered open insofar as it expands on both sides of a lifespan, from the status of future generation before conception and birth to the status of a surviving memory after death; (vi) should human psycho-spiritual inner life also be conceived of as a form of open life in its own right, rather than as a feature of self-centered life – and can this stimulate new philosophical directions in human ethics?

On more technical grounds, one is also lead to explore how the philosophical tradition can integrate this notion of the open character of plants, which at this stage is still more of an aesthetic, or even an "aesthesic", experience than it is an actual thought form shaped by experience, as a pragmatist approach demands. Although mathematical theories provide tools for addressing non-Euclidian properties, the link between theory and actual experience of acting forces remains to be established.

For this reason, I wish to conclude with a plea in favor of the development of an experimental philosophy, a new realism based on interdisciplinary art-science-philosophical interactions as a new mode of enquiry and research action – with the aim of elaborating new shared experiences of the open character of plants.

Acknowledgements

I wish to thank Christine Roberts and the editors for language editing and comments.

Notes

1. Greenness is rather rare outside the plant kingdom. However, green shades can be found in the mineral and animal kingdoms but green animals are usually not green *inside*.
2. Some species classified as animals do not have a bilateral left-right symmetry, for example the sponge family. Biophysicist Vincent Fleury argues that the jellyfish is not an animal but a sort of reproductive part of an animal (Fleury 2006, 177).
3. Aristotle, *Physics*, III 201a10f: motion is defined as the "actuality [entelechy] of the potentially existing *qua* existing potentially" (*hê tou dunamei ontos entelecheia hêi toiouton*). Aristotle uses the words *energeia* and *entelecheia* interchangeably to describe actuality as a kind of action.
4. Aristotle, *Physics*, IV 212a20: a *topos* is defined as the "immediate motionless boundary of what contains" (*to tou periechontos peras akinêton proton)*.

References

Arasse, D. 2004. *Histoires de peintures*. Paris: Denoël.

Aristotle 2002. *Physique*. Paris: Flammarion.

Foster, B. J. 2011. *Marx écologiste*. Paris: Editions Amsterdam.

Bergson, H. 1925. *L'évolution créatrice*. Paris: Librairie Félix Alcan.

Besson, Y. 2009. Une histoire d'exigences: philosophie et agrobiologie. L'actualité de la pensée des fondateurs de l'agriculture biologique pour son développement contemporain". *Revue Innovations Agronomiques* 4 "Agriculture biologique": 329–62.

Baudry, S. 2012. Reclaiming urban space as resistance: the infrapolitics of gardening. *Revue française d'études américaines* 1, no. 131: 32–48.

Brague, R. 1990. Aristotle's definition of motion and its ontological implications. *Graduate Faculty Philosophy Journal* 13, no. 2: 1–22.

Calame, M. 2007. *Une agriculture pour le XXIe siècle. Manifeste pour une agronomie biologique*. Paris: Editions Charles Léopold Mayer.

Callicot, J. B. 1995 Intrinsic value in nature: a metaethical analysis. *Electronic Journal of Analytical Philosophy* 3. http://ejap.louisiana.edu/EJAP/1995.spring/callicott.1995.spring.html.

Chiurugwi, T., Pouteau, S., Nicholls, D., Tooke, F., Ordidge, M. and Battey, N. 2007. Floral meristem indeterminacy depends on flower position and is facilitated by acarpellate gynoecium development in *Impatiens balsamina*. *The New Phytologist* 173, no. 1 (January): 79–90.

Claviler, C. 2009. *Cérès et le laboureur. La construction d'un mythe historique de l'agriculture au XVIIIe siècle*. Paris: Editions du Patrimoine.

Cubas, P. 2004. Floral zygomorphy, the recurring evolution of a successful trait. *BioEssays* 26, no. 11 (November): 1175–184.

Debaise, D. (ed.). 2007. *Vie et expérimentation. Peirce, James, Dewey*. Paris: Vrin.

Delaporte, F. 1979. *Le second règne de la nature: essai sur les questions de la végétalité au XVIIIe siècle*. Paris: Flammarion.

Deleuze, G. and Guattari, F. 1980. *Capitalisme et schizophrénie 2. Mille plateaux*. Paris: Minuit.

Douady, S. and Couder, Y. 1996. Phyllotaxis as a dynamical self organizing process (Parts I, II, III). *Journal of Theoretical Biology* 178, no. 3 (February): 255–312.

Federal Ethics Committee on Non-Human Biotechnology (ECNH). 2008. *The dignity of living beings with regard to plants. Moral consideration of plants for their own sake*. Berne: ECNH.

Fel, L. 2014. Plastique écologique de l'art contemporain: l'exhibition du végétal. Pp. 165–76 in *Le modèle végétal dans l'imaginaire contemporain*, edited by Inès Cazala and Marik Froidefond. Strasbourg: Presses Universitaires de Strasbourg.

Fleury, V. 2006. *De l'œuf à l'éternité. Le sens de l'évolution*. Paris: Flammarion.

Fleury, V. 2009. Clarifying tetrapod embryogenesis, a physicist's point of view. *The European Physical Journal – Applied Physics* 45, no. 3 (March). 30101, doi: 10.1051/epjap/2009033: 1–54.

Foucault, M. 1967. *Des concepts autres*. Repository of texts by Michel Foucault. Accessed from http://foucault.info/documents/heteroTopia/foucault.heteroTopia.fr.htlm.

Fujimori, H., Yano, Y. and Hayashi, Y. 2013. The deconstruction of medicine – Metamorphosis of Chinese traditional medicine in Japan and the birth of morphologic pathology in Edo period. *Journal of Philosophy and Ethics in Health Care and Medicine* 7 (December): 59–80.

Gest, H. 2002. History of the word photosynthesis and evolution of its definition. *Photosynthesis Research* 73, nos. 1–3 (July): 7–10.

Hadot, P. 2004.*Le voile d'Isis. Essai sur l'histoire de l'idée de nature*. Paris: Gallimard.

Hallé, F. 1999. *Eloge de la plante. Pour une nouvelle biologie*. Paris: Seuil.

Henderson, L., D. 2013. *The fourth dimension and non-Euclidean geometry in modern art*. Cambridge, MA: The MIT Press.

Houle, K. 2011. Animal, vegetable, mineral: ethics as extension or becoming? The case of becoming plant. *Journal for Critical Animal Studies* 11, no. 1–2: 89–116.

Houle, K. 2012. "Devenir plante". *Chimères* 76 "Ecosophie": 183–94.

Houle, K. 2014. Les différentes symétries des plantes. *Chimères* 82 "L'herbe": 155–67.

Houle, K. (In press). "Symmetry and asymmetry in conceptual and morphological formations: the difference plant body growth can make to human thought." In *Posthumanisms Through Deleuze*, edited by Daigle, C. and McDonald, T. Bloomington: Indiana University Press.

Houle, K. *Facing only outwards? Plant bodily morphogenesis and ethical conceptual genesis*. This volume.

Kagan, S. 2010.Cultures of sustainability and the aesthetics of the pattern that connects. *Futures* 42, no. 10 (December): 1094–101.

Kallhoff, A. 2014. Plants in ethics: Why flourishing deserves moral respect. *Environmental Values* 23, no. 6 (December): 685–700.
Latour, B. 2015. *Face à Gaïa. Huit conférences sur le nouveau régime climatique*. Paris: La découverte.
Marder, M. 2013. Of plants, and others secrets. *Societies* 3 no. 1 (March): 16–23.
Margulis, L. and Schwartz, K. V. 1982. *Five kingdoms. An illustrated guide to the phyla of life on earth*. San Francisco: W.H. Freeman and Company.
Mayr, E. 1957. Species concepts and definitions. *The American Society for the Advancement of Science* 50: 1–22.
MacLeod, M. and Rubenstein, E. 2006. *Universals*. The Internet Encyclopedia of Philosophy, edited by J. Fieser, and B. Dowden. www.iep.utm.edu/universa/
Molas, M., L. and Kiss, J., Z. 2009. Phototropism and gravitropism in plants. *Advances in Botanical Research* 49 "Artificial photosynthesis": 1–34.
Pimbert, M. 2011. *Participatory research and on-farm management of agricultural biodiversity in Europe*. London: International Institute for Environment and Development.
Poincaré, H. 1902. *La science et l'hypothèse*. Paris: Flammarion.
Pouteau, S. 2012. Providing grounds for agricultural ethics: The wider philosophical significance of plant life integrity. Pp. 154–59 in *Climate change and sustainable development. Ethical perspectives on land use and food production*, edited by T. Potthast and S. Meisch. Wageningen: Wageningen Academic Publishers.
Pouteau, S. 2014. Beyond "second animals": Making sense of plant ethics. *Journal of Agricultural and Environmental Ethics* 27 no. 1 (February): 1–25.
Pouteau, S. 2016a. Des modalités d'intervention "art-science-philosophique" pour éprouver les temporalités de l'urgence environnementale. *VertigO* 16, no.1 (May). Accessed from http://vertigo.revues.org/16976: 1–26.
Pouteau, S. 2016b. Point, ligne et plante: l'être végétal comme expérience de seuil existentiel. Pp. 345–62 in *Les limites du vivant*, edited by R. Barbanti and L. Verner. Paris: Dehors.
Ramade, B. 2007. Mutation écologique de l'art? *Cosmopolitiques* 15 "Esthétique et espace public": 31–42.
Ruse, M. 1995. The species problem. Pp. 171–93 in *Theories and rationality in the biological sciences*, edited by G. Wolters and J. G. Lennox. Konstanz/Pittsburgh: Konstanz Universitätsverlag and University of Pittsburgh Press.
Ryan, J., C. 2012. Passive flora? Reconsidering nature's agency through Human-Plant Studies (HPS). *Societies* 2, no. 3 (September): 101–21.
Skinner, D. J., Hill, A. T. and Gasser, C., S. 2004. Regulation of ovule development. *The Plant Cell* 16, no. 1 (January): 32–45.
Thomas, F. and Bonneuil, C. 2009.*Gènes, pouvoirs et profits. Recherche publique et régimes de production des savoirs de Mendel aux* OGM. Versailles: Quae.
Thuillier, P. 1984. Espace et perspective au Quattrocento. *La Recherche* 160, no. 15 (Novembre): 1384–98.
von Uexküll, J., 2010. *Milieu animal et milieu humain*. Paris: Payot et Rivages.
Wolpert, L. 1992. Gastrulation and the evolution of development. *Development* 116, Supplement (April): 7–13.

8 What do we mean by a relational ethics?

Growing a relational approach to the moral standing of plants, robots, and other non-humans

Mark Coeckelbergh

Introduction: The question concerning "relational" ethics

What does taking a "relational" approach mean in and for plant ethics? Relational ethics has become something of a buzzword recently, but it is not always clear what people mean when they use the term. In addition, the contexts in which the term is used are very different, including for instance care ethics (e.g. Noddings 2012), animal ethics (for an overview see Henderson 2009), robot ethics (Coeckelbergh 2014), and now plant ethics.

This paper clarifies what meanings of "relational" are already in use in plant ethics, and then discusses new meanings of "relational" for plant ethics by drawing connections to robot ethics, a topic of growing importance in the philosophy of technology. The focus is on the issue concerning moral standing: what is the moral standing of plants? Building on previous work in this area (Coeckelbergh 2012, 2014; Coeckelbergh and Gunkel 2014), I question the very ways in which the notion of moral standing is typically understood in ethics, arguing that it is too distant and detached, and that even relational approaches to the ethics of plants such as ecological and network approaches are at risk of taking in too detached a notion of "relational" – at least if they are meant as a metaphysics and insofar as the ethical reasoning they propose is based on the ontological properties of plants.

My own proposal comes in three steps: (1) include the subject in the question, leading to a first-person perspective and drawing attention to my relation to the entity; (2) start from the needs of the other (second-person perspective), and ask not "what kind of moral standing does this entity have?" but "what do you need and what can I do for you?"; and (3) replace the detached model of the viewer observer (which is still implied even in the other-oriented approach) with the model of the active and practical relation to the entity. This focus brings us back to the life world, the world which we live and experience rather than the world as constructed by abstract moral or scientific theory, and a world where the question regarding moral standing actually matters. Finally, I propose the Wittgensteinian concept of "form of life" (Coeckelbergh 2017a, 2017b; Wittgenstein 2009) to think about change in plant ethics. I also argue that the ethical question regarding plants and other non-human entities has a societal and political dimension, and add that art can help us find new possibilities of relating to plants and other non-humans.

Existing relational approaches in plant ethics and their problems

One way of taking a relational approach to plant ethics is to start from a relational view of the world – a relational metaphysics. This could be, for instance, an ecological view that emphasizes that plants, animals, humans, and indeed any other living being are always related to other living beings. Many people working in animal ethics and plant ethics would subscribe to such a view. Plants, for instance, are clearly not isolated "things" but living beings that are part of a larger ecological and relational whole, sometimes called the ecosystem. "Relational" here refers to what we say about the nature of being. It is not entirely clear how this metaphysics should generate an ethics (a typical question here is whether one can ever derive an "ought" from an "is").

In environmental philosophy, an important approach that represents this way of thinking is Deep Ecology (Naess 1973), which relies on metaphysical holism. (Sometimes, deep ecology is also associated with the work of Martin Heidegger, as in Zimmerman (1986); but I will not discuss that view here.) The basic thought behind Deep Ecology is that if we understand that each living entity is related to a living, ecological whole, then we should take into consideration the consequences of our actions on all such living entities out of concern or respect for the whole of which they are parts. This can be extended to plant ethics: one could argue that plants deserve moral standing since they are part of an ecological whole. Whether this means that all plants have a right to live and what, for instance, a no harm principle would imply for plants are different questions. My point here is just that regardless of whatever specific ethical argument one makes from within Deep Ecology, this argument will be based on a relational metaphysics of ecological wholeness.

Similarly, an ethics of technology could argue that moral standing of artifacts such as robots depends on the relational whole in which entities are embedded (Coeckelbergh 2012). So-called "individual" robots are in fact embedded in environments and are related to other technologies. It might be objected that this analogy between plants and robots is inappropriate because robots interact with their environment whereas plants do not. Yet there is scientific evidence that plants can register and adaptively respond to changes in their environments. Consider for instance the literature on plant intelligence (e.g. Mancuso and Viola 2015; Trewavas 2014), which argues that plants are not passive but adapt to their environment and signal opportunities and threats to one another. One could plausibly say that this entails that plants are "relational" or have "relational" properties in a strong sense: not only do they relate to other living beings as beings that are "passively" part of an ecological whole, but they also "actively" relate to those other beings and to that whole as they "perceive" their environments and "communicate" with other beings in these environments. Marder (2012), for example, has written of 'plant intentionality'. Perhaps plants are more like animals than we think, or perhaps they have a special kind of "life-form" of their own (here meant in the biological sense) that is, for example, much more mobile than we think (in fact, whether we want to call plants mobile or not will largely depend on the timeframe that we adopt: by human standards they seem to be very slow). In addition, plants and technologies are related. Influenced by Latour (2005), one could argue that robots are not isolated entities but are part of socio-technological networks or systems. For example, Johnson and Miller (2008) have made such an argument about autonomous computer systems. This can also be applied to plants: plants are not only related to other living beings but they are also embedded in socio-technological networks and systems, including humans but also technologies and infrastructures (e.g. water technologies and infrastructure). Plants are thus always related to humans and non-humans.

On these relational views, the moral standing of plants cannot be defined apart, and receives its relevance from, these larger wholes. Plants as living organisms may surely be viewed as morally relevant, but there is no such thing as "individual" plant life. Living means relating; there is always dependence on other living and non-living entities. In addition, if one believes that from this property of relatedness follows a duty to take care of plants, then the question of how this duty is to be fulfilled will inevitably already be asked from within a specific socio-technological context, which has specific technological means available to take care of plants and particular social rules, games, practices, and meanings available to understand and structure that moral enterprise: for instance the practice of gardening (Di Paola 2017).

Another way of understanding the notion of relational ethics is to point to the relationship that we (as individuals or collectives) may have with a plant, or some ensemble of plants as in a garden – a relationship of care, for example. Just like one can have a relationship with another human being, it may be argued that one can have a relationship with an animal, a plant, a place, a monument, or a robot. What matters here is not so much the metaphysics of wholes or networks but the relationship that we humans – particular humans – have with the plant (and other entities). If there is a relationship, then moral obligations and virtues may be connected to that relationship. For instance, if the plant in the office is my plant, not in the sense of being my property but in the sense that I have a special, personal relationship to it, then I may have a duty of care for it, or it could be said that my character will be shaped by how I treat the plant – depending on the ethical framework used. Here the plant is a different sort of moral patient than in the case of (deep) ecology: it is not only passively or actively part of a whole or a network; it is also part of a relationship, particularly a personal relationship and a relationship of care. In the case of the office plant, I care for the plant as well as the relationship. This may give me a further duty of care, and influence my virtue. This may not be the case for another person who does not have this personal relationship to this particular plant – but who may have other plant relationships. Relationships to plants, in turn, are part of social and cultural practices, such as gardening or forestry. There are already practices, norms, and ways of life that have to do with plants: there is already a culture of plants. For example, I am not the first person ever to take care of an office plant; there is already a practice and a culture. One could also link this approach to an ethics of care more generally (e.g. Gilligan) and to virtue ethics (e.g. MacIntyre, but also in environmental ethics, for example Sandler and Cafaro 2005); but then one would want a virtue ethics that is more inclusive than usual, and involves not just humans but also animals, plants, and other living beings.

More work is needed to discuss these theoretical directions, but the socio-technical, care, and virtue approaches are interesting for my purposes since they already mark some steps towards a less abstract way of thinking about moral standing than do approaches that revolve around a metaphysics of plants (even when the metaphysics is itself relational). These more relational understandings of ethics encourage us to consider the question of the moral standing of plants in relation to other entities, including the human. Let me develop this thought and argue that the question of what moral standing plants (and robots) have should not be asked in the abstract but rather placed in the context of the relation that we have with the entities in question.

Alternative directions, or how to get closer to plants and other entities

The relational approaches presented above are typically minoritarian in philosophical debates on moral standing, which tend to focus instead on the properties of entities. The reasoning often goes as follows:

> Property *p* is sufficient for moral standing *m*
>
> Entity *x* has property *p*
>
> Therefore, entity *x* has moral standing *m*

This is problematic, since it detaches us from the concrete relationships and contexts in which the question of moral standing arises and, worse, from the "patient" itself: as moral observers and reasoners, we take a god's eye point of view, a "view from nowhere" (Nagel 1986). Similarly, when we describe the entity as part of a larger whole, it seems as though we can have an overview of this whole, as though we stood outside or above it. So we end up reasoning about plants as though we had nothing to do with plants.

This detachment is an ethical problem in its own right. Of course abstract moral reasoning may not license harms done to plants; its outcome may be the proclamation of a duty to respect plants, for example. But the point is that the very *procedure* of detached reasoning can create the conditions for moral distancing. As I argued in previous work (Coeckelbergh 2012), only when we take moral distance from an entity can we violate, harm, use, and abuse it. The distance makes the violence possible. For example, if we only talk about animals in the abstract, we can more easily kill them – usually *have* them killed – than if we personally knew particular animals, in the sense of having a personal relationship with them (e.g. with pets). Moreover, by the same procedure, we also turn humans into abstract entities that become unrelated and distant to the entity in question. Thinking that we can be situated outside or above the whole and reasoning as if we were can have severe moral consequences, as it makes us judges of what can and cannot be done to Earth and the beings that inhabit it. If this worry needs to be framed in terms of moral theory at all, one could say that this is problematic from a virtue ethics point of view (and perhaps it could also be framed in deontological and consequentialist terms). But perhaps *theory as such* is also problematic, in so far as it distances us from the entities under consideration.

This problem is especially acute for those approaches that are based on a metaphysical view of the world or a scientific view of the entity. The metaphysician takes an outsider point of view, as though she is not part of the ontological order. The scientist – as scientist – takes a so-called "objective" point of view: this enables her to describe plants, their properties, and their relations; but it also detaches her from concrete plants and the experience we have of them and the relationship we may have with them. How can this problem be remedied?

First, instead of considering the properties of entities (e.g. plants) from an outsider's perspective, the relational ethics that I propose includes the subject in the picture and understands so-called "objective" properties as constructed by our interpretations. Importantly, this move away from ontology (what is) and towards phenomenology (experience) and hermeneutics (interpretation) does not make the moral standing of entities *merely* subjective. The ascription of moral standing is always both subjective and objective, as it is the outcome of the interaction between subject and object. The subject-observer matters to the outcome of the ascription process, but so does the object. What the moral standing of the plant is depends both on me and the plant: it depends on my interpretations and on what interpretations the plant affords as I relate to it. The question, then, is no longer "what moral standing does the entity have?" but "what standing do we ascribe?", "what makes this ascription possible?", and "what are we assuming when we say this and that about the entity?". In this respect, it matters greatly what kind of language we use to talk about plants. For example, do we call a plant a "thing" and use "thing" language, or do we name it as we would a city or a human? How and why do we use the word "weed" with respect to some plants and not others when, biologically speaking, we may in fact be talking about very similar entities?

Elsewhere, I have proposed a transcendental argument to conceptualize this approach (Coeckelbergh 2012) – with "transcendental" understood not in the abstract sense that typically belongs to aesthetics and metaphysics but in a more material and practical sense: we must critically investigate and evaluate the conditions of possibility of our ascriptions of moral standing. These conditions can be linguistic, technological, and cultural at the same time. For example, we may consider how gardening practices, crop-breeding technologies, ways of speaking about plants, and the rules and norms embedded in our culture turn certain plants into weeds. The ascription of moral standing is the outcome of a process that engages all these tools and media. The question that is relevant for plant ethics then is no longer what the plant "is" and whether that justifies ascriptions of moral standing, but how we "make up" its moral standing. This enables us to take a critical perspective on how we relate to plants – for instance, on how we "make" particular plants into "weeds" (undesirable and expendable) by means of our language and practices.

This first step is still characterized by a first-person perspective, which we incorporate but also surpass with a second step. In the lifeworld and in practice, we do not only observe plants or talk about them from a distance and in the abstract. We also actively relate to them – when we eat them, for instance, or when we come to look at and treat them as companions. We can thus try to further reduce moral distance by taking on a second-person perspective: the perspective of "the other", as Lévinas (1969) dubbed it. Whether the other is a plant, a robot, a human being or any other entity, we start from the needs of the other and the "face" of the other as we meet it in concrete encounters (i.e. we start from the demands and the appeal of the other), and what we ask is not "what kind of moral standing does this entity have?" but "what do you need, and what can I do for you?". For Lévinas, this approach applies only to humans. But we can also apply it to other entities, for example animals (Coeckelbergh and Gunkel 2014). At least *potentially*, animals can be others, and so can plants. For plant ethics, then, the question is if plants can appear as others and, if so, how to respond to the plant as other.

This other-directed approach further helps our exploration of our relationship with a plant and its ethical dimension. Is the plant a mere means for the formation of my character, as some virtue ethics approaches may suggest, or is the plant totally other, as a Lévinasian account expanded beyond the human would imply? And is the plant similar to me in important ways, or is it radically different? Most likely, we will want to find middle ground with respect to both these questions. The plant should not be a mere means for my character formation: the good life should include the flourishing of both me and the plant. And the plant is not totally other: it is a living being like me and has similar needs – it also needs water, for instance. Concerning many of my needs, I can surely relate to plants. However, insofar as they are different and remain other, perhaps I cannot use plants merely as means and I cannot reduce them to, say, general categories of biology or metaphysics.

In Lévinas there is an encounter with the other, but this still entails some distance. I have an encounter with the plant, but it remains alien to me, as though there was no further relationship, no history, no narrative that connects us. Therefore, I propose that we take a third step and replace the model of the viewer-observer with that of the active and practical relation to the entity, a relation which has a history and a narrative of lived experience. What matters then, ethically speaking, is not only a particular "encounter" or "situation" as such but our on-going engagement with the entity: how we actively and practically relate to it, and the ethical possibilities that arise out of this relation. The question, then, is not "what moral standing does this entity have?" but "what can I do with you?", "what needs to be done now in regards to you?", and similar. What matters is not a particular response at a particular moment in time (the encounter) but, rather, how to shape, understand and live a larger practical relation with the entity. This relation has a history and a narrative, and we can consider the whole of our relation to the other, rather than a specific aspect or moment of it, in light of this history and narrative. This opens up questions on how to treat a particular plant, or robot or any other entity, what skills and know-how are needed (Coeckelbergh 2015) and, because we are reflecting on how to continue and shape relations that have histories and narratives that overlap with our own, generally on how we live our lives and think of ourselves.

The detached perspective of the moral philosopher and observer is thus replaced by that of the active practitioner and sense-maker: we find ourselves in the lifeworld, understood as a practical and relational space where the question of what to do for and with a plant matters. In this lifeworld, notice, there is not only an "I" but also a "we". Plant ethics is also a matter of asking "what shall we – as a collective – do with plants?". The question how to relate to plants and other non-humans is also a political, societal and collective question. Many theories in ethics presuppose an individual decision maker, but there are also decisions that need to be made as a community and as a society. This requires an ethics that involves specific ideas about communication and participation. (This aspect could be further developed by discussion various philosophical frameworks, including, for instance, virtue ethics and communitarianism.)

This move to morality-in-practice and morality in the lifeworld and society does not mean that we simply immerse ourselves in the everyday, without further reflection. If the proposed approach to moral standing is still to be ethical, philosophical, and critical, it needs to take *some* distance. By using language, we can take some distance and question our practices. As said, this can be done by revealing, constructing, and critically questioning the conditions of possibility of those practical relations and their contexts. We can question our current relations to plants by critically examining the language that we use when we talk about them (and perhaps also to them). Proposals by philosophers to talk about plants in different ways, for instance as persons (Hall 2011) rather than things, can be interpreted and developed as interventions in, and comments on, not only our uses of language but also the activities, practices, and rules that are connected to these uses of language.

One example of language use that works as a condition of possibility for our thinking about plants is our use of thing-language, as already noted. Another is our very use of the word "plant". Just as Derrida (2008) questioned the word "animal", we can question the word "plant", which seems to be too generalizing and possibly incongruous with what we presently know about plants. For instance, the knowledge that plants seem to be a kind of very slow-moving *animal* (see for instance plant scientist Jack Schultz's claim in a recent interview (Gabbatiss 2017)) can lead to new uses of language, new explorations of how to best talk about plants. Or one may choose to better define what kinds of plants one is talking about, both in terms of biology and their relation to us. In addition, the term "plant" is ill defined (it is not clear, for instance, where the boundary lies between bacteria and plants, or indeed between plants and animals) and also lumps together an enormous diversity of life forms and relations. One may also add that such a term is itself a tool for the exclusion of other entities: perhaps it is true also of the term "plant" what Derrida (2008, 34) said of the term "animal", namely that it denotes "all the living things that man does not recognize as his fellows, neighbors, or his brothers". It is also important to remember that we should not only look at how philosophers use language to speak about plants: we must also study how all of us, in concrete practical contexts, speak about (and possibly to) plants – for instance during gardening, eating, and outdoor activities (or indeed in the lab, in the office, and so on). Here, philosophers may want to collaborate with the social and natural sciences, working in a trans-disciplinary way.

Turning to concrete practical contexts does not mean the end of ethics or moral philosophy. On the contrary, attending to our moral language, for instance, may lead to new conceptual innovations and a more, rather than less, critical attitude. Perhaps philosophers should use their concepts differently: not to say something about the nature of being, for example of plant being, but rather to say something about us-and-plants, about what we do with plants and what plants do with and to us, and about the history, the present, and the future of human–plants relations. Morality itself has a history: to develop a plant ethics, it is important to also understand that history, and why it has led morality to say this or that about plants. We need to know more about how plants are already treated and related to, and reconstruct what makes these treatments and relations possible. If we neglect all this, these histories and practices and relations will feed and structure our thinking about plants anyway; we better reveal them and reflect on them if we want our thinking and relations to plants to be better informed and more critical. This entails that changes in the ways we relate to and think of plants are path-dependent and thus cannot be changed overnight. After all, once we radically question or even abandon abstract and detached moral reasoning, we'll find ourselves with no external criteria to hang on to. We'll be criticizing our own form of life (e.g. with respect to plants) from within this very form of life.

Forms of life: The "standing" and place of plants and humans in the Anthropocene

In his *Philosophical Investigations*, Wittgenstein used the concept of 'form of life' to express the idea that that our particular uses of language and their meanings are embedded within a larger whole that structures that meaning – our "form of life" (Wittgenstein 2009). Before we learn to speak there is already a language and there already are ways to use words. These ways grow out from and in turn reinforce the ways in which we do things in our times, society and culture.

The problem regarding distance and abstraction discussed above can be reformulated as a problem that concerns language use. The problem is that we talk about plants and other entities as though they are mainly and merely objects (or moral patients) from which we can easily distance ourselves. As philosophers we use language in certain ways to ascribe a standing to entities. These language uses are part of our activities and form of life, as participants in practices and in a culture that include specific ways of relating to plants, for example scientific and philosophical ways. These larger ways of talking and doing already shapes our thinking about plants, and make it difficult to conceive of radically different ways of thinking and talking about them. Our thinking about plants is shaped by specific activities, practices, and forms of life, such as eating and gardening, that bring with them specific ways of talking about plants and relating to them. Language use, thought, and practical ways of relating to plants are all connected and reciprocally reinforcing.

When we talk about plants and their moral standing, we speak from a position where we have already integrated them within our own human "form of life" and within particular forms of life in certain places, contexts, and practices. As our lives are shaped in specific ways as humans and as members of social-technological wholes, we already relate to plants in certain ways and not others. Questioning our relation to plants, then, also involves questioning ourselves, as persons but also participants in a culture and as humans: it involves, perhaps necessitates, questioning our own form of life. The moral standing of plants crucially depends on their standing within our activities and our form of life. Their standing depends on *where they stand* and, ultimately, on where *we* stand. The question about the moral standing of plants is, for a relational ethics of the sort that I have been describing, also a question about our standing as humans. Where do we stand in relation to plants?

If we turn the question of moral standing, initially understood as a question concerning moral ontology, into a question concerning moral ecology and moral geography, we end up with an approach that transforms the enterprise of examining the moral standing of plants into the enterprise of examining how *I* relate to plants and *we* relate to plants and, in the end, how we *humans*, as relational beings, relate and are related to in general. Moreover, forms of life always change and so will the moral standing of plants, since the entire relational constellation of which both plants and humans are parts will also change. In this sense, plants or other entities have no "standing"; what they are and do is not fixed. This counsels the adoption of an historical perspective: to critically reflect on their moral standing, we have to look at the history of plant-human relations. It opens up the inquiry to several other disciplines beyond philosophy, including anthropology and psychology. But even trans-disciplinary work within academia may not be enough for a thorough re-thinking of human–plant relations.

We will, perhaps, need artists or coders to help us explore new possibilities for relating to plants and, more generally, the natural environment.

Finally, the notion of "form of life" also enables us to link our inquiry on plants to human flourishing. We do not start from a naturalistic or metaphysical view of the properties of plants and their life-form, and link that to the notion of flourishing (Kallhoff 2014); we rather emphasize plants and humans as part of a social-cultural form of life. We can then ask how the flourishing of the plant as a life-form is related to human flourishing, particularly human flourishing as it is intrinsically connected to a social-cultural form of life.

At first glance, this link between human flourishing and plant flourishing only makes sense for plants that are directly integrated in human societies and practices, such as garden plants and plants used for food production (agricultural plants). However, in the context of what is currently often referred to as "the Anthropocene", a notion which is used to stress that human activities have now global geophysical impact on the planet that we inhabit (see for example Steffen et al. 2011), and that the future of humanity and technology is one with the future of the planet, it becomes increasingly difficult to consider plants or anything/anyone else on Earth as unconnected to the human. This means, once again, that to re-think the moral standing of plants is to re-think our entire form of life, which is not only human but also non-human at the same time, involving plants, animals, and all kinds of technologies. In the Anthropocene these human and non-human, natural and cultural, plant and animal forms of life are always and fatally entangled with one another.

Conclusion: Deconstructing plants (and humans)

What is a relational approach to the moral standing of plants? What could such an approach mean for the ethics of plants and related fields? In this chapter, I have first identified some common ways in which the notion of "relational ethics" can be or has been used in plant ethics, particularly in accounts that rest on a relational metaphysics such as Deep Ecology or in accounts that are "relational" in the sense of concerning "relationships". To these existing accounts I have added some understandings of relationality that change the very question of moral standing.

By bringing in notions like alterity, lifeworld and form of life, I have reframed the question of the moral standing of plants and other non-humans as one that concerns our practical and lived relations – not only as individuals but also as collectives, communities, societies – to and with plants. I have argued that pursuing this question must include questioning our uses of language and other conditions that make possible and shape these concrete human–plant relations. The question regarding the moral standing of plants and other non-humans is thus not only about plants but also about us, humans. It is about our life and virtue as persons, about the ways we do things in our societies and cultures, and about the future of humanity and the planet.

To re-think our relation to plants, then, is also to re-think our form of life. And given how dependent our human form of life is on plants, animals, the rest of nature, and technology, this re-thinking includes questioning the nature and future of the whole: a whole hybrid human/non-human form of life, which usually functions as an unquestioned background and "grammar" for our actions, our relationships, and our lives. This is especially urgent in the Anthropocene, when it becomes even clearer that our human fate is deeply entangled with that of plants and other life-forms, and with the future of technology.

The challenge for a philosophical ethics that aims to be "relational", in plant ethics and elsewhere, is now to further develop conceptual tools that may help us critically reflect on the conditions of possibility that shape and structure our relations to plants and other entities. Questioning the very word "plant", as a gesture of deconstruction, is a good start. Questioning the word "human" may have to follow.

Acknowledgements

I warmly thank all colleagues and speakers of the workshop "Plants and Robots" (Vienna 2016) for the great discussions we had. I am also grateful to Marcello Di Paola and Maria Schörgenhumer for their editorial work.

References

Coeckelbergh, M. 2012. *Growing moral relations: Critique of moral status ascription*. Basingstoke/New York: Palgrave Macmillan.

Coeckelbergh, M. 2014. The moral standing of machines: Towards a relational and non-Cartesian moral hermeneutics. *Philosophy & Technology* 27, no. 1: 61–77.

Coeckelbergh, M. 2015. *Environmental skill: Motivation, knowledge, and the possibility of a non-romantic environmental ethics*. New York: Taylor & Francis.

Coeckelbergh, M. 2017a. Technology games: Using Wittgenstein for understanding and evaluating technology. *Science and Engineering Ethics* (online first) Available from https://link.springer.com/article/10.1007/s11948-017-9953-8.

Coeckelbergh, M. 2017b. *Using words and things: Language and philosophy of technology*. New York and Abingdon: Routledge.

Coeckelbergh, M. and Gunkel, D. J. 2014. Facing animals: A relational, other-oriented approach to moral standing. *Journal of Agricultural and Environmental Ethics* 27, no. 5: 715–733.

Derrida, J., Mallet, M. L., and Wills, D. 2008. *The animal that therefore I am*. New York: Fordham University Press.

Di Paola, M. 2017. *Ethics and politics of the built environment: Gardens of the anthropocene*. Dordrecht: Springer.

Gabbatiss, J. 2017. *Plants can see, hear, and smell – and respond*. Available from www.bbc.com/earth/story/20170109-plants-can-see-hear-and-smell-and-respond.

Hall, M. 2011. *Plants as persons: A philosophical botany*. Albany, NY: State University of New York Press.

Henderson, K. 2012. *Relational approaches to animal ethics and the intuition of differential obligations* (Master thesis), University of Victoria.

Johnson, D. G. and Miller, K. 2008. Un-making artificial moral agents. *Ethics and Information Technology* 10, no. 2–3: 122–133.

Kallhoff, A. 2014. Plants in ethics: Why flourishing deserves moral respect. *Environmental Values* 23, no. 6: 685–700.

Latour, B. 2005. *Reassembling the social: An introduction to actor-network-theory – An introduction to actor-network-theory*. Oxford: OUP Oxford.

Lévinas, E. 1969. *Totality and infinity: An essay on exteriority*. Pittsburgh, PA: Duquesne University Press.

Mancuso, S. and Viola, A. 2015. *Brilliant green: The surprising history and science of plant intelligence*, transl. by J. Benham. Washington/Covelo/London: Island Press.

Marder, M. 2012. Plant intentionality and the phenomenological framework of plant intelligence. *Plant Signaling & Behavior* 7, no. 11: 1365–1372.

Naess, A. 1973. The shallow and the deep, long-range ecology movement. A summary. *Inquiry* 16, no. 1–4: 95–100.

Nagel, T. 1986. *The view from nowhere*. Oxford: Oxford University Press.

Noddings, N. 2012. The language of care ethics. *Knowledge Quest* 40, no. 5: 52–56.

Sandler, R. and Cafaro, P. 2005. *Environmental Virtue Ethics*. New York: Rowman and Littlefield.

Steffen, W. et al. 2011. The anthropocene: From global change to planetary stewardship. *Ambio* 40, no. 7: 739–761.

Trewavas, A. 2014. *Plant behaviour and intelligence*. Oxford: Oxford University Press.

Wittgenstein, L. 2009. *Philosophical investigations*. Malden: Wiley.

Zimmerman, M. E. 1986. Implications of Heidegger's thought for deep ecology. *The Modern Schoolman* 64, no. 1: 19–43.

9 Caring for plants

Cultivating relational virtues

Maria Schörgenhumer

Introduction

Among the manifold ways of relating to plants, care stands out as a practice that focuses on promoting a plant's own good. While care is usually framed as a virtue in environmental ethics, often without further explanation, this chapter takes a somewhat different approach: a close consideration of care shows that it is more than just a simple virtue; rather, it can be understood as a relational practice – a way of relating to and dealing with plants – that demands specific virtues. Since caring for plants is an essential part of gardening, I will be repeatedly referring to this context in the course of this chapter, which addresses the following questions: What is care? What does it mean to care for a plant? What kind of evaluations and normative aspects are involved? How is care linked with plant flourishing? Which virtues are needed so that care does not fail? And why should we care about caring for plants?

What is care?

In environmental debates, care is mentioned from time to time as a preferable attitude towards nature or natural entities. In her analysis of virtues in the ecological discourse, Louke van Wensveen names care as one of the "core ecological virtues" (van Wensveen 2000, 118). Roger King points to the "relevance of the vocabulary of an ethics of care to ecofeminism" (King 1991, 75). As for plant ethics, care has been explicitly mentioned by Matthew Hall, who finds it in human–plant relationships that are established in restoration – in contrast to the instrumental relationships he suspects behind traditional ways of cultivating nature, like gardening (Hall 2011, 157–69). Nevertheless, care towards plants can also be found as a virtue in the ethics of gardening, where it is the plant's own good that guides the gardener's actions (Brook 2010, 16–24; Cooper 2006, 94–95; Kallhoff and Schörgenhumer 2017, 203–4), and it can even be argued the garden provides an especially fertile soil for care.[1] But first, what is care? And what does it mean to care for plants?

In discussing care, one distinction is important right from the beginning: *to care about* or *to care for*. *To care about* means that somebody/something is important to the caring subject, whereas the stronger notion of *caring for*, which I shall focus on, does not only indicate a committed relation of holding somebody/something dear, but also the practical engagement of protecting, helping, fostering the other, of promoting the other's good. Following this active, interactive sense of care, I conceive of *caring for plants* as a relational practice. Given the lack of plant-specific explanations of care, my inquiry starts out from a more general explanation, drawing on the well-elaborated analysis of care by Nel Noddings (2013; 2011, 41–46), which shall then be transferred to a people–plant setting.

To begin with a basic definition, to care (for) is to "act [. . .] to protect or enhance the welfare of the cared-for" (Noddings 2013, 24), or, a little more detailed (ibid.): "The one-caring desires the well-being of the cared-for and acts (or abstains from acting – makes an internal act of commitment) to promote that well-being." First of all, this involves being attentive and sensitive to the other in order to properly read the needs expressed by them, followed by a "motivational displacement" (Noddings 2011, 42): putting aside one's own interests and wanting what is best for the other. Yet, just having the other's best interests at heart is not enough. As the key verb in the provided definitions says, it is also required to take action. To care for somebody means to *do* something in favor of that person, to respond to their needs so as to protect and enhance their well-being. This may or may not work, depending on how well A understands B's needs, and whether A's intentions and caring actions turn out to be appropriate and beneficial for B; anyhow, B reacts to A's care in some way, and this feedback allows A to evaluate and improve their caring for B. Noddings emphasizes that B must contribute to the caring relation at least by some form of positive response in order to hold up A's motivation to care and to stay "committed to maintaining or improving the caring relation itself" (Noddings 2011, 42).

So, in a successful caring relation, the part of the cared-for basically consists in expressing their states of being, their needs as well as their reactions to the care received. The part of the one-caring requires reading this "caring information" correctly, and continuously responding to it by doing the right things to promote the other's good.

Caring for plants

Care as explicated above is not restricted to interpersonal relationships; it also makes sense when addressed to non-human living beings, for (or insofar) they too are capable of flourishing, hence vulnerable, and capable of expressing needs or, more generally, displaying their states of being in a "readable" way. (If there were nothing to "read" and no kind of possible well-being on the part of the other, care would be pointless.) Persons can care for animals, persons can care for plants. Furthermore, care can also be applied to larger ensembles of living things, whole biotopes or cultivated nature like "the lawn" or "the garden." Flourishing can be attributed to them too; it is bound to the flourishing of the plants they consist of. Yet, caring for a garden does not mean to generally care for all the plants involved, but to selectively care for *some* plants that are wanted there, while unwanted plants are pushed back or eliminated (e.g. "weeds").

In the following, I refer to caring for specific plants, or even to caring for a plant in singular, since this is the form of caring which targets and promotes the good of a plant as a living being in its own right.[2]

As a possible objection to applying the concept of care to plants, Noddings highlights the one-sidedness of people–plant relationships: while human beings are able to care, plants are not, which means that the roles of the carer and the cared-for could not possibly be reversed. But then again, interpersonal caring relationships are, in most cases, also asymmetric; a baby child cannot care for its mother, a sick or elderly person in need of help cannot take the carer's part. This doesn't change, not even challenge the validity of the concept of care as applied to plants. Moreover, the people–plant relation is not that one-sided, for plants do respond to a person's care in different ways. Indeed, as Noddings also remarks, plants are "responsive cared-fors" (Noddings 2013, 159–60).

Considering the basic requirements of care as discussed in the first section, it can be stated that plants can be cared for in the following sense:

1. Plants have needs and those needs can be read. A plant's state of being – and, over time, its development – is expressed through its body. A good gardener knows what his plants need: not only by theoretical/practical knowledge and experience, but also by looking at them. Key to "reading" plants is the aesthetics of flourishing, as discussed in the following section.
2. One can actually do something to promote a plant's "well-being" and positive development. One can *foster* it by providing whatever it needs, enhancing its living conditions (soil quality/nutrients, water, light, temperature/humidity, space to grow), even supporting its proliferation. One can further *protect* its flourishing: from harmful weather, pests, or competing plants that threaten to overgrow it, but also from human action/disregard like being trampled, rammed, cut down, or being hurt in other ways.
3. Plants are sensitive to what (environmentally) happens to them. In this respect, they are also sensitive to care, and they respond to it – if all goes well, by obvious flourishing.

Aesthetics of flourishing – Responsive care

According to Angela Kallhoff, plant flourishing can be defined by three aspects: (a) a plant's vitality and capability of coping with external stress, (b) the accomplishment of its typical life cycle, including proliferation, and (c) the expression of typical plant/species features (Kallhoff 2014, 687). This is a theoretical, abstract concept of flourishing that is useful for making normative claims in plant ethics; however, it does not fully cover the sense of flourishing that underlies the *practice* of plants care, the concrete "concept" of flourishing a gardener works with – beyond conceptual and moral reasoning. When we are faced with a live plant, its capability of coping with stress (a) is something we cannot perceive, and it most certainly has not completed its full life cycle yet (b), for it is still living, growing, maybe in its prime of life. Still we can tell if it is flourishing: we can see that. Hence, in order to explain the practice-immanent evaluations of plants care, I propose an aesthetic account of flourishing. Engaged in the practice of caring for plants, a gardener evaluates *in situ* how the plants are and what they need; caring choices are not only based on knowledge (including central aspects of the theoretical concept of flourishing),[3] but essentially also on how a plant looks like at a certain state in the process of development. In aesthetical terms, flourishing is something that leaps to the eye.[4]

A flourishing plant is beautiful: lush and green, without discolorations, abundant in leaves, buds, blossom, fruit; both the whole plant body and its parts are unfolding symmetrically and proportionally (as measured by the idea of an "ideal plant" of the species in question), without disruptions of their "gestalt", their visible integrity; the plant is growing upright and holding up its body parts in good tissue tension. In all of this, the plant displays strength, vitality, abundance – unbroken beauty.

If a plant's flourishing is somehow impaired, this again is evident in its appearance: in its diminished or broken beauty. "Rusts, smuts, powdery mildews, downy mildews, galls, leaf curls, scabs, necrotic spots, cankers, diebacks, blights, wilts, rots, yellows, mottles and mosaics, color breaks, phyllody" are all symptoms of plant illnesses (Ingram, Vince-Prue and Gregory 2008, 226) and at the same time aesthetic irritations. Such signs of impaired flourishing – the list could be completed by pest infestations and injuries of plants – are aesthetically perceived as ugly and disturbing, as pointed out by Karl Rosenkranz in his *Aesthetics of Ugliness* (Rosenkranz 1853, 18). They lead the gardener to take action: to make them go away – either by curing the plant, removing its ugly/ill parts, providing it with water, nutrients or whatever it may need in order to get better and well again, or, if it is too late, by removing the whole plant from the garden. A good gardener realizes small signs in time and prevents setbacks in flourishing from the beginning.

It can be stated that a plant, through its aesthetic appearance, communicates its state of being and its needs in an intuitively understandable way. This is the basis for responsive care: the caring gardener responds to what is expressed by a plant, and is encouraged and rewarded by the beauty of flourishing, the plant's "response" to good care. This kind of feedback both informs and motivates further care and hereby contributes to "maintaining or improving the caring relation itself" (Noddings 2011, 42).

"But do they [plants] respond to my caring or merely my caretaking?" (Noddings 2013, 159–60) – Although plants don't know the difference, it can be argued that a person that truly cares for their plants does the job better than a person that just takes care of them by performing some actions required by protocol (ibid.). Care involves being attentive to the other's needs and responding to them as appropriately and well as possible, whereas "caretaking" means following some externalized rules like "water the plant once a week" without taking notice of the plant's development under the given conditions (it may very well be that the plant, given its age, state of development, and environmental conditions, needs more or less water, or something else).[5]

Failures and virtues of caring for plants

In my understanding, caring for plants is a relational practice that requires specific virtues, and by seriously engaging in such practice, these virtues can be learned and deepened.[6] A plant (or whatever living being) can be cared for poorly or well; one can be good or bad at caring, and one can improve one's caring skills, extend one's knowledge, and learn virtues that are essential to good care. So what does it take to *care well* for plants? To explain the content of care thoroughly and to get a hold of the virtues in question, it is instructive to clarify how care can fail.

Broadly speaking, care fails when it misses its aim: the flourishing of the cared-for. In the most obvious case, this happens when a caring relation is cancelled and a gardener just stops caring for his plants (although continuous care would be necessary), be it because he has given up and just "doesn't care anymore" (lack of persistence and commitment), or be it because it has become impossible for him to carry on (e.g. for time or health reasons, or a lack of resources). In any event, this is different from the situation where someone refrains from active care for the sake of the cared-for, when it is best to let the other be and develop at its own pace, for its own good (on "letting be" as a virtue of gardening cf. Cooper 2006, 158–60), in which case the caring relation is maintained.

Care can also be misguided. One can do a lot of things with intent to promote a plant's flourishing and still no good might come from it. The most important reason for this is a *lack of attention*:

> A does not listen to B or only half-listens; that is, she [. . .] recognizes bodily signs, but she reads these through a preselected cognitive scheme. She assesses or evaluates or diagnoses B's condition and infers a need, and she then acts to satisfy the inferred need – one which may be entirely different from the one expressed by B.
>
> (Noddings 2013, 42)

Without true openness for what the other communicates, without closely "listening" to a plant, one can go wrong in doing too much, or too little, or just the wrong things compared to what would be best for the plant, because one does not know its actual needs. A gardener might still know what is required in theory, or from experience, but then again, this can also block the view; by blindly proceeding by the book, the actual needs of a plant "here and now" are likely to be ignored and overridden. Strictly speaking, without turning one's attention to the other, one does not even begin to care: no caring relation is established where a person takes no real interest in how the other is.[7] Lacking attention is hence the principal failure of care. In other words, attentive openness for the other is the basis that underlies and constitutes the relation of true care.

Furthermore, care can also fail due to a *lack of understanding/knowledge*. Plant signals can be read attentively, yet wrong (misinterpretation), which leads to wrong caring decisions. Besides, not knowing or being mistaken in assuming what is best in this or that situation for this or that plant under specific circumstances also leads to wrong decisions. Nevertheless it has to be considered that in dealing with evolving living beings one can never know everything in advance. A good part of care is experimental; caring has much to do with getting familiar with and learning what is best for the cared-for, which involves experimenting and, again, demands openness for learning about the other. A gardener would only have to be accused of real ignorance if he refused to learn something new, stubbornly repeating his mistakes again and again, never improving his care.

Impatience is another failure of caring for plants. Flourishing takes its time, it cannot be enforced, and in order to support it properly, a gardener needs to respect natural time cycles, especially a plant's own time of development through the times of year and over the years. Acting too soon (e.g. sowing, planting out, cutting, harvesting) is counterproductive to flourishing.[8]

Finally, care can also fail due to a *lack of commitment*. A gardener might see a plant in need and still not do anything about it, maybe because it would be tedious, or difficult, or dirty, or exhausting, or because there just are better things to do. While it is true that many acts of caring are tedious, it is all the more important to be committed to promoting *the good of the other*, to put aside one's own interests in favor of the cared-for (this corresponds with what Noddings calls "motivational displacement" (Noddings 2011, 42), a central aspect of *caring for*). At worst, a lack of commitment results in a cancelled caring relation, as explained in the beginning of this section.

Drawing on the previous reflections on failures of care as well as its key aspects that have been explained in the sections before, it now just remains to list the virtues that have turned out to be important for caring well for plants. They are necessary for a good practice of plants care, and we learn them *by doing*, so it can be said that caring for plants entails or also "induces"[9] these virtues: (1) attentiveness and openness for the other, (2) a commitment to the flourishing of the other, (2) vigor in the sense of getting things done, taking action to promote the other's own good, (4) patience and persistence, and (5) a willingness to learn, to improve care by growing understanding and knowledge, which is more important than "knowing everything" and, different from the latter, essential to caring.[10] Finally, there is (6) humility to be learned, which indicates knowing one's limits, especially regarding nature (also see Kallhoff and Schörgenhumer 2017, 202–3).[11] Of course, there are deeply human aspirations tied to keeping gardens and cultivating plants. Yet, the practice of caring for plants also requires and teaches us to acknowledge the limits that nature puts to our human endeavors. For one, in caring for a plant we need to respect its autonomy, its own time of development, its own needs and vulnerabilities, its own terms of flourishing. And at the same time we need to accept what is fundamentally unavailable in nature more generally: natural processes of growth and decay that affect, or rather pertain to, all life, including human life. The most optimized and virtuous care is not powerful enough to guarantee flourishing, and a beloved cared-for might suffer and die even if we do anything in our power to prevent that.

Why caring for plants matters

Caring for plants is an important way to realize relational virtues and to pay respect to plants as flourishing beings. However, it is also relevant beyond that.

First, there are many settings in which people do personally care for plants – in gardening, especially in private and botanic gardens, in small-scale farming/kitchen gardening, in fostering house plants –, and these practices are valuable because they contribute to a good life. Caring for plants brings us in contact and keeps us concerned with the beauty and otherness of plant life. In caring for plants, we learn about the vulnerability as well as the amazing vitality and capacities of plants, and we are rewarded by the beauty of flourishing, or even by the literally nourishing fruit of our labor (Schörgenhumer 2016).

Furthermore, care is a good way of relating to plants, and to nature more generally, because it requires – as well as it induces and improves – knowledge, skills and virtues that are both personally and environmentally valuable. The virtues of caring improve a person's relationship to plants, and to nature more generally; alongside with further virtues of gardening, they constitute good practices of dealing with nature, which makes them important *environmental virtues* (see Kallhoff and Schörgenhumer 2017). As Isis Brook states, "gardening improves both people and land" (Brook 2010, 13), and the same is true in a narrower sense for caring, which is at the core of gardening. Caring for plants not only promotes their flourishing, it also involves improving the soil and the environment they are embedded in, and it "improves" the caring person in that she learns and practices the virtues that care demands.

Admittedly, a plant does not recognize the efforts of human care, it just reacts to "environmental conditions favorable or unfavorable to the realization of its good" (Taylor 1981, 210), and plants can very well flourish without people caring for them, but this does not mean that care is obsolete. There is an ever growing range of plant species on Earth that could not survive any more, or at all, without our help. This accounts for both wild plants and cultivated plants. Endangered wild species threatened of extinction are cared for in botanic gardens. And cultivated plants are dependent on human care from the beginning; their evolution, survival and flourishing is bound to cultivation. They need special growing conditions, protection against competing plants (weeds) and other "enemies" in nature (animals, pests, bad weather), and in many cases also help with reproduction, for the ways of proliferation are often altered through plant breeding.[12] So, while it is neither necessary nor appropriate nor possible to care for *all* plants, many plants are in need of care, and if these plants are valuable to us (for whatever reasons), we need to care for them.

Notes

1. Matthew Hall also claims a close connection between gardens and care; he mentions care as an important gift that persons give to plants; care "helps to build significant relationships between humans and plants" – the basis of "ethical behavior towards the cultivated plants in the garden" (Hall 2010, 45).

2. I presuppose that a plant's own good is intrinsically linked with its flourishing (Kallhoff 2014).
3. Knowledge about plants, or more specifically, about the typical features a plant should display according to its species (a), about a plant's typical development/life cycle (b) and its individual "biography", refined by experiences in cultivating this kind of plant under the specific local conditions, is crucial for reading a plant correctly and making the right decisions.
4. What is left aside here, yet important to keep in mind: the perception of flourishing is, of course, not only visual, but involves different senses; haptic and olfactory impressions also play a significant part in determining a plant's flourishing.
5. This marks an important difference between personal care as in small-scale gardening (and hands-on agriculture) on the one side, and industrialized plant production on a large scale on the other side, where people are no longer in direct contact with the plants and alienated from nature by machines or even replaced by robots in fully automatized farming settings, rendering human involvement virtually obsolete. In these styles of plant cultivation, not only the personal relation to nature, and with it, a great deal of understanding, is lost, but also the aesthetic dimension of flourishing, which is both communicative and rewarding, goes missing when no one even looks at the plants any more.
6. Note that this understanding differs from approaches in environmental ethics and the philosophy of gardening where care is defined as a virtue (as mentioned in the beginning). Yet, it seems too complex to be reduced to a simple virtue, or a virtue at all (cf. Noddings 2007, 42). In contrast to the virtue sense of caring, Noddings describes care as a *relation* to which she attributes particular virtues, first of all "attention, motivational displacement, and response" (ibid.). I largely agree with this account, however, as argued earlier, I propose a different accentuation by suggesting that caring for plants be understood as a *relational practice*. Noddings also mentions other virtues of caring, but most of them are linked with interpersonal communication and cannot be applied to a people–plant setting; only patience and "(striving for) competence in the area of need" (Noddings 2007, 42) also make sense regarding plants care.
7. Even if one starts out from just closely observing a plant continuously over a period of time, this constitutes a relation that goes beyond objectification and recognizes the other as a "center of life" (which could be considered the basis for care), as Paul Taylor describes: "As one becomes more and more familiar with the organism, one becomes fully sensitive to the particular way it is living out its life cycle. One may [. . .] even experience some involvement with its good and bad fortunes (that is, with the occurrence of environmental conditions favourable or unfavourable to the realization of its good)" (Taylor 1981, 210).
8. Patience is hence an important virtue of gardening (Brook 2010, 20; Cooper 2006, 96; Harrison 2008, 78).
9. On virtues being "induced", in this case, by garden practices, cf. Cooper (2006, 91).
10. Noddings also emphasizes this aspect: „*striving* for competence in the area of need" (Noddings 2007, 47).
11. Cooper describes humility in this sense as a kind of "unselfing" (Cooper 2006, 94–97) that is learned in dealing with nature, as in gardening. "Seeing our human plans and fancies overridden by natural processes in the garden" (Brook 2010, 21) is an essential experience of gardening that induces humility in the sense of knowing one's limits regarding nature (also see Čapek 2010; Pollan 1991).
12. For the differences between wild plants and cultivated plants, as brought about by cultivation, see, for example, Brücher 1977, 25.

References

Brook, I. 2010. The virtues of gardening. Pp. 13–25 in *Gardening – Philosophy for everyone. Cultivating wisdom*, edited by D. O'Brien. Oxford: Wiley-Blackwell, 13–25.
Brücher, H. 1977. *Tropische Nutzpflanzen. Ursprung, Evolution und Domestikation*. Berlin: Springer.
Čapek, K. 2010. *Das Jahr des Gärtners*. Berlin: Aufbau-Verlag.
Cooper, D. 2006. *A Philosophy of Gardens*. Oxford: Oxford University Press.
Hall, M. 2010. Escaping Eden: Plant ethics in a gardener's world. Pp. 38–47 in *Gardening: Philosophy for everyone – Cultivating wisdom*, edited by D. O'Brian. Oxford: Wiley-Blackwell.
Hall, M. 2011. *Plants as persons. A philosophical botany*. Albany, GA: SUNY Press.
Harrison, R. P. 2008. *Gardens: An essay on the human condition*. Chicago, IL: The University of Chicago Press.
Ingram, D., Vince-Prue, D., Gregory, P. (eds.). 2008. *Science and the garden: The scientific basis of horticultural practice*, 2nd ed. Oxford: Blackwell.
Kallhoff, A. 2014. Plants in ethics: Why flourishing deserves moral respect. *Environmental Values* 23: 685–700.
Kallhoff, A. and Schörgenhumer, M. 2017. The virtues of gardening: A relational account of environmental virtues. *Environmental Ethics* 39, no. 2: 193–210.
King, R. 1991. Caring about nature: Feminist ethics and the environment. *Hypatia* 6, no 1 (Spring): 75–89.
Noddings, N. 2011. Caring as relation and virtue in teaching. Pp. 41–60 in *Working Virtue: Virtue ethics and contemporary moral problems*, edited by R. Walker and P. Ivanhoe. Oxford: Clarendon Press.
Noddings, N. 2013. *Caring. A relational approach to ethics and moral education*. 2nd ed. Berkeley, CA: University of California Press.
Pollan, M. 1991. *Second nature: A gardener's education*. New York: Grove Press.
Rosenkranz, K. 1853. *Ästhetik des Hässlichen*. Königsberg: Bornträger.
Schörgenhumer, M. 2016. Good home-grown food? A relational approach to kitchen gardening. Pp. 547–61 in *Food futures: Ethics, science and culture*, edited by Olsson, ed.. Wageningen: Wageningen Academic Publishers, 547–61.
Taylor, P. 1981. The ethics of respect for nature. *Environmental Ethics* 3, no. 3: 197–218.
Van Wensveen, L. 2000. *Dirty virtues: The emergence of ecological virtue ethics*. Amherst, New York: Humanity Books.

Part II

Appreciations and applications

10 Forest ethics

Robin Attfield

My aim in this paper is to contribute to plant ethics through contributing to the area of forest ethics. I will assume that the contours of plant ethics are already well established and understood (Kallhoff 2014, 685–700), but before proceeding to ethical conclusions about forests, I should explain the nature of the field of forest ethics itself.

The main reason for this deviation from the path of straight-forward plant ethics is that many and perhaps most plants live, thrive and die in large assemblages. To consider the benefits and the harms that befall plants and how human beings should either treat them, seek to increase their numbers, make use of them, or decide to leave them alone, we need to consider these assemblages or, in other words, thickets, woods or, to mention the main assemblages of trees, forests. Forests, however, are more than the sum of their constituent plants and more than the sum of their constituent living members, and forests exercise an influence that these individual members would not have if they lived in the same numbers in isolation. For example, forests act as watersheds, absorbing rainfall, and in some cases as stimulants of rainfall or other forms of precipitation: in these and other ways, forests modify the climate of the lands in which they are situated. They also supply habitats for reproduction and growth, not only for the species native to the forest but often for others as well. Forests have historically been refuges, during periods when the savannah temporarily contracted, where large concentrations of biodiversity were often preserved in relatively small localities.

In these and in other ways, forests determine the character of landscapes, both those where they remain widespread and, to some extent, those that they used to occupy but from which they have been largely cleared, as occurred in much of Europe between the Middle Ages and the present. Often, after having been cleared at one time, forests are either replanted (as in France) or themselves recolonize the areas taken from them and become the characteristic background scenery of places such as the temples of Angkor Wat in Cambodia and the holy places of Machu Picchu in Peru.

Forest ethics includes the consideration of all those areas of human activity that intersect with woods and forests, including the justification, and rightful extent and character, of such activities. Not all plant ethics are part of forest ethics because some issues, such as the well-being of plants, are relevant both to forest plants and other, non-forest plants, such as the plants of gardens, fields, rivers, lakes, botanical gardens, and parks. Additionally, not all forest ethics constitute a branch of plant ethics: some branches of forest ethics concern the deployment of forest products (such as rubber and brazil nuts) and the preservation of animal forest species. However, for the most part, the issues that characterize forest ethics, such as whether forests are to be considered as resources, and if so, whether they are to be considered as nothing but resources, are also central to plant ethics. This being so, we can confidently delve into forest ethics, and even touch on the related issues raised in the ethics of forestry, in the confident hope that the current address is unlikely to be entirely irrelevant to a series of essays on plant ethics.

Many different issues are associated with the felling and uprooting of trees in addition to the undermining of the value of particular existing trees. These different issues often need to be considered together if defensible overall judgments on issues such as deforestation and reforestation are to be reached.

I have argued for a biocentric stance in ethics in the past (Attfield [1983] 1991, [1981]1994, 35–54), and I continue to adhere to such a stance, although I recognize that far from everyone is in agreement about this. Given that trees have a good of their own, and that this good has intrinsic value, biocentrists hold that, similar to other bearers of such value, trees and their interests should be considered in decision-making. However, if we biocentrists are consistent, this approach involves not only concern for individual trees, some of which are in danger of being burned or felled, but also consideration of future trees and the assurance that there will be ecological niches for them to grow and flourish. Indeed, non-biocentrists should also consider these matters, albeit out of concern for human and animal interests rather than those of plants. The felling of forests, however, and their replacement by streets and houses or by fields and farming, means that no forest trees can grow in these places in the future until the streets or the fields fall into disuse after many centuries, as has occasionally occurred in places where nature has taken back land, replacing villages, towns or farmlands.

Thus, the removal of forests should be of even greater concern than the felling of the same number of trees and should certainly be of greater concern than cases where the felled trees are replaced by others. (Indeed, the felling of trees in a plantation and their replacement by others will not always be ethically objectionable at all, or so I hope to show later.) However, let us next remind ourselves of the damage or the harms arising from the extermination or obliteration of a forest.

One of the most common consequences of the destruction of a forest is the release of carbon dioxide into the atmosphere. Such releases of carbon dioxide will not always be due to the unmitigated destruction of forests since the felled trees may be used as timber, retaining most of their carbon. However, large proportions of felled trees are usually either burned or allowed to rot, which, in both cases, leads to releases of the carbon dioxide that had been captured and retained by the tree with little or no prospect of this carbon dioxide being reabsorbed by the vegetation growing in place of the trees felled. Such releases add to the proportion of carbon dioxide in the atmosphere. Historical levels of atmospheric carbon dioxide were 280 parts per million by volume, which is the level that persisted, more or less, for several centuries before the industrial revolution. Recently, the level of atmospheric carbon dioxide has been reaching 400 parts per million. As the scientists of the Intergovernmental Panel on Climate Change have been warning, such increases threaten global warming, rising sea-levels, increased flooding, an increase in severe weather events, and associated problems such as the uprooting of millions of environmental refugees; the spread of diseases such as malaria and dengue fever to places of higher altitude and higher latitudes (as their vectors spread away from the equator and towards the poles), and the disruption of species, both animal and vegetable, as their habitats disappear (Houghton 2004).

Many of these threats are admittedly threats to humanity and to human interests. However, biocentrists can be concerned for human well-being (since human beings are among the beings with a value of their own) and can additionally be concerned about the disruption to species that are often forced to migrate towards the poles until they run out of habitat at the limits of the lands that they used to occupy. Therefore, biocentrists actually have an ampler basis for concern about climate change than anthropocentrists or sentientists have; anthropocentrists are concerned for humans and their value, sentientists are concerned for sentient animals (including humans) and their value, but neither of them is concerned for plants as such.

In addition to its contribution to climate change and the problems that attend it, deforestation also contributes to aesthetic and biodiversity losses. Once again, it is not the value of the relevant individual trees that is the main focus where aesthetic loss is concerned; instead, it is the human love of trees, woods and forests, and the related sorrow and distress when trees or woods are destroyed, that is the focus. Fostering a love of trees is foreseeably central to fostering motivations for preserving them.

Such love of trees can be encouraged through visits to woodlands for a variety of reasons, including to exercise, recreate, escape the streets of towns and cities, find asylum and experience peace and quiet, or to conduct biological studies. Love of trees can also be fostered through art and poetry; think of the song about the Linden Tree in Franz Schubert's *Winterreise* or Gerard Manley Hopkins' poem 'Binsey Poplars' ('Felled, felled, all felled'), which has encouraged many to deplore the felling of trees, particularly historically significant trees. Deforestation deprives people of the opportunity to appreciate the trees and woodlands that their predecessors used to enjoy. Some trees are landmarks and thus carry symbolic value for the human beings who are familiar with them; but if biocentrists are correct, all flourishing trees carry intrinsic value and there are grounds to regret their demise, even in cases where, in light of other considerations, their felling is justified.

We should also bear in mind that deforestation contributes to biodiversity loss. Species are being lost at a more alarming rate than ever before, and this unprecedented loss is largely ascribable to human activity. Particular concentrations of species are (or used to be) found on coral reefs and also, more relevantly, in rain-forests, which are widely at risk (respectively) because of practices such as indiscriminate fishing and forest clearing for the sake of agriculture and mining. Since it is forests that are under consideration here, we should bear in mind that in many places, rain-forests have unparalleled concentrations not only of trees, flowering plants, epiphytes and funguses but also of insects, spiders, fishes, amphibians, birds, reptiles and mammals. Ecological communities, including representatives of all or most of these categories of life, form (together with inanimate entities such as rocks and soils) ecosystems that, once disrupted, cannot easily be reassembled if they can be reassembled at all. Other kinds of forests, such as temperate and boreal forests, have their own ecosystems but are seldom as rich and prolific in species as tropical forests, where many species are confined to small localities and disappear forever if the forest of those localities is cleared.

In view of all these considerations, it might be supposed that the practice of forestry, or the deliberate rearing of trees with a view to the production of products such as timber and paper, is unjustifiable. Or, if it is justifiable based on anthropocentric or sentientist ethics, it might be supposed that this practice should at least be condemned by biocentrists. How, it might be asked, can anyone who recognizes the moral standing of trees or the far-reaching harms generated by cutting them down be prepared to counter this arboricidal practice, let alone benefit from it?

There are certainly some limits to the moral acceptability of forestry. Certain particular practices, such as the clear-felling of natural forests, should usually be condemned as unjustified even by consistent anthropocentrists who take the long view of human interests and include the interests of future generations, which, if the clear-felling of whole forests occurs in the present generation, will be deprived of all possibility of studying and appreciating such forests in their own day. Even if selective felling is justifiable because it is a practice that facilitates regrowth, clear-felling reduces the prospect of regrowth too severely to be acceptable.

This does not, however, mean that all forms of the practice of forestry are to be condemned. When forestry practices are being appraised, the impacts to the benefits and harms for humans need to be considered, as well as those for domesticated animals, other nonhuman animals, and plants, including trees. In addition, when benefits and harms to humans are considered, the importance of timber, particularly as a building material, needs to be kept in mind. Indeed, it is arguable that human beings do and will always need timber to satisfy basic human needs.

By now, enough has been said to show that the felling of trees answers genuine and widespread human needs that are unlikely ever to be otherwise provided for. Hence, the value of satisfying these needs has to be weighed against the other values at stake, including the value of the trees liable to be lost to forestry, and the value of the other inhabitants of forests both vegetable (whether natural or planted by humans), and animal, including farmed animals that need housing that often requires the use of forest products. In addition, while the needs of human beings and domesticated animals will seldom warrant the clear-cutting of natural forests, the case for rearing trees in plantations that are then felled for timber, paper or related products, is much stronger.

It seems particularly hard to resist the view that trees are, in the ordinary course of events, replaceable. While there are cases of landmark trees with historical or symbolic value that gives these particular trees the status of irreplaceability, the value of most trees can be retrieved or recovered after they have been felled, through the planting and rearing of other trees. Peter Singer has argued for the replaceability of conscious creatures (such as cows) that are not self-conscious (Singer 1993); the absence of an on-going sense of selfhood on the part of these creatures means that it is not wrong to kill them painlessly when other similar creatures will be raised and reared in their stead. It is the killing of self-conscious creatures that he argues is normally wrong, because these creatures are thus deprived of the futures of their expectations and in many cases of their own devising. Anyone sympathetic to Singer's reasoning is likely to agree that trees, which hardly anyone regards as conscious, let alone self-conscious, can be regarded as replaceable. In addition, even those who regard conscious, as well as self-conscious, creatures as irreplaceable are unlikely to take this view of non-conscious beings such as trees.

This does not mean that there is no loss of value when trees are felled or otherwise killed. There are aesthetic losses to human beings from the loss of their beauty, as well as symbolic losses. There are also losses related to the release of the carbon that had been sequestered in the tree when a tree is felled, which is something that probably occurs to some degree whenever felling occurs. There are also losses to other creatures, such as to the birds who used to nest in the trees, the insects that used to make their homes in their trunks and branches, and the parasites such as mistletoe and fungi that similarly depended on trees' hospitality. In addition, biocentrists can add that the intrinsic value of the tree, or of its flourishing, is lost as well. All these values normally supply a good case for leaving trees alone and standing.

However, given the needs of human beings and domesticated animals, there is also a strong case for selective tree-felling, particularly if the felled trees are likely to be replaced. Given the more complex and sophisticated nature of these needs, biocentrists, as well as others, can recognize this, since there is no requirement for biocentrists either to disregard human and other animal interests or to treat all living creatures as having equal moral significance. (Some biocentrists such as Paul Taylor do seem inclined to take this latter view (Taylor 1986), but they usually pull back from its practical implications. In any case, there is no need for biocentrists to be strictly egalitarian in matters of the moral significance they ascribe to different creatures.) Biocentrists are likely to be in the forefront of those seeking to limit the deforestation of natural forests and seeking to limit the human use of forest products, but biocentrists have no need to oppose forestry plantations as such.

We should also refrain from disregarding the needs for fuel-wood of millions of people living in developing countries. Hundreds of millions of people in the

Third World have no supply of electricity (Hildyard, Lohmann and Sexton 2012, 2013) and need fuel both for cooking and for domestic heating. Unlike the human needs mentioned previously, this need may be hoped to be transitory, since if society becomes better organized, then electricity can increasingly be supplied to most of these people – through, for example, hydroelectric schemes, which do not exacerbate the problem of climate change. There is a developmental case to be made for developing countries to invest in such schemes and for developed countries to assist them. Indeed, biocentrists would be consistent in adding their support, since schemes of this kind would tend to reduce the need for the felling of trees.

However, since the practice of foraging for fuel-wood tends to clear the afforested areas close to cities, we should also uphold schemes that support the deliberate rearing of plants for fuel-wood and the selling of that fuel-wood inexpensively to those who need it. The plants concerned can then once again be regarded as replaceable, and it will be important to replace them, at least as long as the need for fuel-wood continues, although this case will become weaker at the stage when sufficient electric supplies become available. The practice of planting plantations for fuel-wood is itself a form of forestry and should, therefore, be regarded as included in the subject matter of forest ethics. My conclusion is that forest ethics should not condemn this practice, but should broadly welcome it with a view to satisfying the human needs of poor people in developing countries.

However, at least equally important is the management of existing forests. Forests can be regarded as renewable resources, although as refuges and repositories of a large proportion of the species on the planet, they cannot be regarded as resources only. However, their renewability makes a large difference to how they can and should be managed. Renewable resources can be harvested recurrently as long as sufficient forest is left to renew itself through re-growth. As long as sufficient forest is left for regrowth and disasters such as hurricanes do not strike, then the forest will be able to renew itself in perpetuity, making it possible for human beings to benefit from it year after year. This is the basis of 'sustainable management', which should, however, also respect both the concentrations of local wild species and the need for creatures such as mammals to have intact corridors of forest through which they can forage or migrate, for example along river banks. The need to respect concentrations of wild species, and particularly species that are endangered, has often involved declaring an area a national park, which is exempt from forestry activities altogether.

Unfortunately, such sustainable management is far from standard practice in the contemporary world. Our Common Future, the Brundtland Report of 1987, relates that 93 per cent of the original primary forest of Madagascar had been eliminated by 1987, and almost all of the forests of western Ecuador had been destroyed to make way for banana plantations, oil wells and human settlements with the loss of an estimated 50,000 species in the 25 years between 1960 and 1985 alone (World Commission on Environment and Development 1987, 149). Similar stories can be repeated for many developing countries with forests, given their pursuit of economic development.

However, as Susan George (1992) has explained, it is not only domestic pressures that have led to deforestation. A leading factor has been international debt and the imperative for indebted countries to pay off at least the interest, if not also the capital. George notes that all sixteen of the most heavily indebted countries with forests in their territories are among the big deforesters, and that five of the eight most heavily indebted countries (Brazil, India, Indonesia, Mexico and Nigeria) figure in the top ten of countries that engage in deforestation. In addition, rates of deforestation accelerated through much of the same period as that of the intensification of debt, that is, from the seventies to the nineties of the last century (George 1992, 10–11). Thus, it is likely that the pressures from creditor countries and from international institutions such as the World Bank and the International Monetary Fund have played a large part in fostering deforestation over recent decades. Other relevant pressures include the economic demand of western consumers for products raised in areas of former forest, such as beef and aluminium, which are produced on lands cleared of forest in Brazil. In any case, deforestation has continued across the period since George drew these matters to attention in the 1990s, and now, most tropical lowland forests are under threat of destruction around the globe.

There is, in any case, a strong case for debt cancellation for heavily indebted developing countries (George 2010, 232–35). Some amount of debt cancellation was achieved through the summit held at Gleneagles in Scotland in 2005; however, much debt remains together with the perverse incentives that it generates to sacrifice forests for the sake of redeeming the debt. Accordingly, forest preservation has become as important a project as sustainable forest management.

To cite a particular case, Tony Juniper relates that to prevent the deforestation and mining of forest areas in Guyana, Norway agreed in 2009 to pay $250 million to Guyana to keep its forests intact (2 years after President Jagdeo had requested such assistance from other European countries). This figure was arrived at partly by calculating the economic value of the ecosystem services provided by these forests, which are jointly larger than England and Scotland combined. The project has been carried through to execution; the money has been used for development and explicitly embodies a model that is 'globally replicable' (Juniper 2013, 72–74). Juniper also relates that in face of the incentives offered by Norway, among other countries, Brazil and Indonesia have reduced their rates of deforestation in recent years (Juniper 2013,166, 289).

The prospects for forest biodiversity preservation, indeed, are not entirely bleak. The protocol generated at the Nagoya summit, held in Japan in 2010, presented possibilities for enhanced international cooperation in preserving biodiversity-rich zones, including zones in developing countries, and having developed counties provide the funding for such preservation. Similarly, the New York Declaration on Forests of 2014 commits many states to forest restoration, including commitments by Ethiopia to reforest 15 per cent of its territory (Suding et al. 2015, 638–40). The execution of these agreements depends on the willingness of the sovereign parties to the international agreement to deliver what they undertook, but here too, there are promising signs. For instance, Indonesia, Brunei and Malaysia have agreed on measures to conserve the forests of Borneo (possibly just in time to save the orangutans that depend on the continued existence of those forests); and in 2013, Bolivia, Brazil, Colombia, Ecuador, Guyana, Peru, Suriname and Venezuela formed the Amazon Cooperation Treaty Organization (ACTO) to promote sustainable development in the context of the UN Forum on Forest (International Institute for Sustainable Development 2013). Such South–South cooperation is every bit as vital as funding from the developed countries of the North.

Another kind of deal for saving forests consists of debt-for-nature exchanges or swaps, which were once supported by the writer and campaigner Susan George (George 1992, 30–31). In such exchanges, a western NGO (non-governmental organization) agrees to pay some of the debt of an afforested developing country with the understanding that specified areas of forest will be preserved from schemes such as forest clearing, road building and mineral exploitation.

Nicole Hassoun has recently argued that debt-for-nature swaps are wrong in circumstances where they infringe on the human right to be immune from extreme poverty. This could, in theory, occur where the foregone development could have met the needs of poor local people, for example for employment. Hassoun adopts a cosmopolitan account of obligations to uphold rights, which include developed states having an obligation to protect the rights of people in developing countries, where the developing countries themselves are unable to uphold the rights of their own population and thus to discharge their obligation to their own citizens. Hassoun concludes that debt-for-nature swaps often breach this obligation, and further argues that poverty and environmental degradation are frequently closely associated and correlated, so that agents will seldom need to choose between them. (Hassoun 2012, 359–77). However, this leaves unaddressed possible cases where we can either save the habitat on which a species depends (e.g. the Borneo forest, home of the orangutans) or protect the human rights of the local people. But often species can go nowhere else than their original habitat. Hence, while schemes should imaginatively address human rights, human rights observance cannot be the sole criterion. Nor are human rights and their observance invariably a necessary condition of the permissibility of preservationist schemes.

This brings us back to the stance Susan George takes in her recent book *Whose Crisis, Whose Future?* (2010). There is now, she claims, 'a dazzling opportunity to reduce the social hardships of the South, provide gainful employment for its people and simultaneously strike a very large blow for the environment' (and in fact for forests). Her proposal consists of an international agreement to cancel the debts of 'every country prepared to comply with certain conditions. . . . In exchange for debt write-offs, for at least X per cent of the debt relief granted, these countries would undertake important reforestation and biodiversity conservation works.' (George 2010, 232–33). This conservation work would preserve forests, contribute to the sequestration of greenhouse gases, and preserve some of the endangered forest species in the spirit of the Nagoya protocol.

Some might accuse this proposal of neo-colonialism: does it not undermine the sovereignty of the developing nations in question over part of their own territory? However, there are possible replies to this objection. The form of the reforestation and the biodiversity conservation work could be dependent on the decisions of the country in question and could also be based on consultation with local people, whose indigenous knowledge of their own forests often proves superior to that of outsiders. With these conditions built into the scheme, a great deal of good could be accomplished both for the local people, not least in terms of employment, and for their forests and the species native to those forests. The exploitation of any mineral deposits could, at the same time, be curtailed, which is consistent with the importance, recently agreed upon by climate scientists, of keeping most of the known deposits of oil, gas and coal in the ground, which is necessary to confine the rise of average temperatures to 2 degrees (Celsius). These being some of the benefits of George's proposal, I conclude that we ought to support this international version of a debt-for-nature exchange.

We have travelled a long way from the plant ethics of individual plants and trees. Plant ethics lead to forest ethics, and it transpires that forest ethics involves a complex background of international debt, biodiversity loss and climate change, as well as widespread social problems situated in developing countries. As we have seen, all these matters are inter-related, and forest ethics cannot be addressed if any of them are ignored. Another conclusion, then, is that plant ethics should often be conducted in as holistic a context as possible with reference to the needs of the societies, particularly poor and developing communities, where trees are situated, as well as the needs of animals and trees. This is fully consistent with a biocentric ethic, which, in my view, handles these matters best, but non-biocentrists can fully subscribe to this kind of development of plant ethics alongside those of a biocentrist persuasion, because everyone, and therefore all ethicists, needs to reflect on the present and future of our forests.

References

Attfield, R. 1983. *The ethics of environmental concern*. Oxford: Columbia University Press.

Attfield, R. 1991. *The ethics of environmental concern*. Oxford: University of Georgia Press.

Attfield, R. 1994. *Environmental philosophy: Principles and prospects*. Brookfield, VT: Ashgate.

Attfield, R. 2001. The good of trees. In *Environmental ethics and philisophy*, edited by J. O'Neill, R. Kerry Turner and I. J. Bateman. Cheltenham: Edward Elgar, Pp. 328–47.

Attfield, R. 2015. *Ethics of the global environment*. Edinburgh: Edinburgh University Press.

George, S. 1992. *The debt boomerang*. London: Pluto Press.

George, S. 2010. *Whose crisis, whose future?* Cambridge, UK and Malden, MA: Polity Press.

Hassoun, N. 2012. The problem of debt for nature swaps from a human rights perspective. *Journal of Applied Philosophy* 29, no. 4: 359–77.

Hildyard, N., Lohman, L. and Sexton, S. 2012. *Energy security: For whom? For what?* Sturmister Newton: The Corner House.

Hildyard, N., Lohman, L. and Sexton, S. 2013. *Energy alternatives: Surveying the territory*. Sturmister Newton: The Corner House.

Houghton, J. 2004. *Global warming: The complete briefing*. Cambridge: Cambridge University Press.

International Institute for Sustainable Development. 2013. Biodiversity policy & practice. Accessed April 25, 2013 from http://biodiversity-l.iisd.org/

Juniper, T. 2013. *What has nature ever done for us? How money really does grow on trees*. London: Profile Books.

Kallhoff, A. 2014. Plants in ethics: Why flourishing deserves moral respect. *Environmental Values* 23, no. 6: 685–700.

Singer, P. 1993. *Practical ethics*. Cambridge: Cambridge University Press.

Suding, K., Eric, H., Margaret, P., Baird, J. C., Christopher, B. A., Matthew, B., John, J. G., Kelly, L. H., Matthew, C. L. and Brendon, M. H. L. 2015. Committing to ecological restoration. *Science* 348, no. 6235: 638–40.

Taylor, P. W. 1986. *Respect for nature: A theory of environmental ethics*. Princeton, NJ: Princeton University Press.

World Commission on Environment and Development. 1987. *Our common future*. Oxford and New York: Oxford University Press.

11 On the relationships between agriculture and landscape

Paolo D'Angelo

The subject of this essay is not the influence of agriculture on the landscape, as its title could suggest. A topic of this sort could hardly be discussed in general terms, especially by a scholar of philosophy. The landscape transformations brought about by agriculture, particularly in countries home to ancient civilizations such as European countries, are so extensive, far-reaching, and firmly entrenched that illustrating them requires painstaking investigation and in-depth competences. In countries like Italy, for example, agriculture has contributed to shaping, organizing and transforming the landscape for many centuries (Sereni 1961).

The topic of this chapter is much narrower: it concerns not the alterations made to the actual landscape, but those alterations that have taken place in our own attitude towards nature and the landscape. I will outline two central trends, taking place at different times.

I will show how for a long time the nature that was loved, perceived as agreeable, and hence appreciated within the landscape was a nature developed by humans, the object of agriculture or at any rate of human labour – in other words, the cultivated countryside. Broadly speaking (and leaving aside certain antecedents to which I will later return) it was only over the course of the eighteenth century, that wild, inhospitable and hostile nature came to be appreciated. Over the last two centuries this idea of wilderness has become the dominant paradigm for natural beauty as a whole. The kinds of landscapes to be admired have been identified with those less affected by human intervention.

Only in recent times – over the last couple of decades have we witnessed a second, reverse trend: a rediscovery of the value of the cultivated countryside as a landscape, restoring its centrality in relation to our perception of natural beauty in general. It would not be far from the truth to argue, then, that while it took us two millennia to develop a love for the wilderness, we have only been following the inverse path, back to a love for the cultivated countryside, for a few years.

This paper will examine first the appreciation of cultivated land in the antiquity e Middle Ages (parts 1 and 2), then the rise of love for uncultivated nature and its triumph in the Romantic Age (part 3); finally, the rise of a renewed appreciation of agricultural environments in our days (final part).

The appreciation of cultivated land in the antiquity

Greek and Roman antiquity harbored suspicion and repulsion towards the wilderness, whilst being aware of its charm. The issue of landscape perception in Antiquity might be discussed at length, and many different opinions have been expressed on the matter, starting from J. Ritter and A. Berque's theses that the notion of landscape is essentially a modern one (Berque 1995; Ritter 1963), and the opposite views held by M. Venturi Ferriolo (2002). Certainly, the ancients possessed a keen sense of space and of what we may describe as a *feeling* for nature, as shown by the connections they constructed between given places and myths, by the very establishment of temples, sanctuaries and oracular sites in highly evocative places, and – in Rome at least – by the arrangement of space for military or urban purposes (Baldo and Cazuffi 2013).

In antiquity natural beauty was detected either in nature *as a whole* or in individual natural beings (e.g. in the human body), rather than in a specific, concrete aspect of nature, as it may be suggested by talk of ancient landscape sensitivity. What is highly revealing, in this respect, is the almost total lack of individualizing representations of places, either in the figurative arts or in literature and poetry. What is most commonly found are rather stereotypical depictions of abstract places, such as rural environments in Theocritean poetry and the representation of ideal landscapes in Hellenistic and Roman painting, for instance in the so-called Odysseus series now in Vatican or in some frescoes in Pompeii.

Now, if we keep to the level of stereotyped descriptions, it is possible to identify an underlying opposition, in antiquity, between the *locus amoenus* (roughly, the lovely place) on the one hand, and the *locus horridus* (roughly, the dreadful place) on the other. This amounts to a contrast between an environment congenial to human life and often shaped by humans, and an environment hostile to human life and impervious to humans.

A pleasant environment may take the form of a verdant meadow strewn with flowers, rich in running water, and offering travelers the cooling shelter of shady trees. An example would be the spot on the shores of the Ilisos, where Socrates and Phaedrus meet in the Platonic dialogue that takes Phaedrus' name. By contrast, a *locus horridus* is marked by a lack of vegetation, reflecting the aridness and sterility of its soil, by vastness and a lack of points of reference – as in Lucan's description of the Libyan desert.

No doubt, the *locus amoenus* is not always a cultivated place. However, it is an idyllic rural and bucolic setting, typically inhabited by shepherds. In this respect, the *saltus* that these shepherds inhabit is distinguished from the *silva* (the forest). The former is a congenial place, the latter is perceived as alien and dangerous. Alongside the pastoral landscape we find the cultivated field (the *ager*) and the garden (*hortus*), which were the ancient Romans' favored natural settings. For the Romans the best vantage point for the observation of nature was provided by the country *villa*, the rural dwelling of wealthy citizens.

The perception of agricultural space is always associated with the perception of the concrete activities that take place within it, what we would call the agricultural industries, as in Horace's celebrated ode:

> That corner of the world smiles for me beyond all others, where the honey yields not to Himettus, and the olive vies with green Venafrum, where Jupiter vouchsafes long springs and winters mild, and where Aulon, dear to fertile Bacchus, envies not the clusters of Falernum. That place and its blessed heights summon thee and me; there shalt thou bedew with affection's tear the warm ashes by thy poet friend!
>
> (Horace, Odes, II, 6)

An antecedent of the modern view of the landscape may be found in Pliny the Younger's description of the environs of a country villa at Tifernum Tiberinum. The author here stresses the beauty of the place, speaking of "regionis forma pulcherrima" (Letters, V, 6) In the writing of agricultural theorists from Varro to Columella, considerations regarding the fertility of the soil and high yield of agricultural estates go hand in hand with an acknowledgement of their beauty as an added value, as it were. When having to choose between two equally productive estates, one should opt for the most beautiful one, since *utilitas* and *voluptas* must not be separated and, most importantly, they should never be thought of as mutually exclusive. As Emilio Sereni has noted, "in Varro, aesthetic requirements coincide with rational and utilitarian ones" (Sereni 1961, 60).

Another typical feature of the ancients' outlook on nature that deserves notice is the link often drawn between inhospitable areas and faraway places, particularly ones inhabited by enemy peoples. Examples are the interior of Anatolia which provides the setting for Xenophon's *Anabasis*, the German forests described by Tacitus, and the wilderness of Caledonia that Hadrian chose to cut off from colonized Britain.

Appreciation of nature in the Middle Ages

Representations of open natural spaces are rare in the Middle Ages. What is relatively common instead, especially from the twelfth century onwards, are depictions of agricultural labor, particularly depictions concerned with the so-called "cycle of the months", for instance the cycle sculpted by Benedetto Antelami in Parma. In these representations natural space is often reduced to a minimum and almost allegorized through the inclusion of an ear of wheat or vine shoot, as in the sculptural calendar adorning the so-called "Porta della Pescheria" of the Cathedral in Modena.

Moving closer to the modern age, however, and directing our gaze to Northern Europe, we can begin to catch a glimpse of some landscapes. For instance, the representation of the month of February in *Les très riches heures du Duc de Berry*, an illuminated manuscript from the early fifteenth century, now at the Condé Museum in Chantilly, in northern France, offers a view of snow-covered hills under an overcast sky and of a valley dotted with village rooftops. To be sure, what stands in the foreground are agricultural tools, a sheep pen and women huddling around a fireplace, whereas the stark forest on the right is shown in relation to the woodcutter who is collecting wood for the fire. In other pictures of the same manuscript the background only consists in a single building and its walls, as in the depiction of springtime haymaking and ploughing.

In Italy, the most famous – and almost unparalleled – instance of the representation of a territory in relation to the agricultural work performed within it is no doubt the large fresco painted by Ambrogio Lorenzetti in the Palazzo Pubblico in Siena in 1338–1339 to illustrate the effects of Good Government. Here too we find a broad view of a hilly landscape. A procession of knights makes its way through the walls of Siena, as a country dweller moves in their direction and other farmers carry produce into the city on mules. In the foreground, reapers are scything hay, while other men are busy harvesting wheat. In the distance, rows of vines already dot the hills. The co-presence of agricultural tasks typical of different seasons clearly betrays the allegorical character of the scene, which, after all, does not illustrate any identifiable stretch of the Sienese countryside.

What we have, then, is not genuine landscape painting: this only emerged in the West one century later in relation to experiences of a different sort, not primarily related to the representation of the cultivated countryside. Thus Van Eyck's famous *Madonna of Chancellor Rolin* (about 1435) offers the view of a river winding its way through forest and city; Antonello da Messina's *Crucifixion* in Sibiu (about 1460) clearly shows the gulf and harbor of Messina in the background of Mount Golgotha with the three crosses; and the imaginary landscapes by Patinier (about 1483–1524) are all fanciful ones made up of dense forests, crags and caves.

Indeed, if the prototype of modern landscape perception is to be found in Petrarch's (1304–1374) description of his ascent of Mont Ventoux, as suggested by Burckhardt (1860), then in this landscape we see the very opposite of cultivated farmland. Petrarch ascends the mountain against the advice of a shepherd, who warns him that only thorns and stones, sweat and toil are awaiting him. The emphasis is on the wild and inhospitable nature of this place, a high mountain that offers nothing agreeable to humans.

What emerges, then, is the contrast between a feeling of nature that for centuries was destined to remain the prerequisite of a tiny fraction of the population – to be found among educated people –, and common man's perception of nature. Petrarch does not provide the only example of love of the mountains, of an environment not marked by human labor and indeed hostile to the presence of humans. The Swiss Humanist Konrad Gessner loved the mountains as well, and devoted a short book to the subject, *De montium admiratione* (1541). Similarly, painted landscapes often feature, if not high mountains, at any rate a glimpse of semi-wild nature.

Things are rather different in the case of the common man: for many centuries still, travelers and writers continued to show appreciation only of the nature that had been made productive by man. In his *Journal de Voyage*, written in the late sixteenth century, De Montaigne warmly describes the beauty of the Po Valley. Almost two centuries later, Charles de Brosses waxes lyrical over the same landscape – "the land extending between Vicenza and Padua alone is probably worth the whole journey through Italy. No art scene is more beautiful and embellished than such a countryside" (De Brosses 1858, I, 153).

The kind of landscape that elicited admiration from, and was contemplated with most pleasure by, the majority of people was the cultivated plane, not the inhospitable mountain landscape. At the same time, the horror of the wilderness and the fear of threatening places endured. These feelings gave rise to popular legends about "accursed" mountains home to monstrous creatures. A traveler such as John Evelyn, describing in his Diary his journey through Italy in 1646, spoke of the Alps as nothing but a rubbish dump in which nature had piled up all the filth and horrors from the plains. Particularly revealing, in this respect, is the curious geological theory developed by Thomas Burnet, the author of *Telluris theoria sacra* (1681), who posited that the Earth was originally flat but then was corrugated, creating the mountains, as a divine punishment.

It was only in the late seventeenth and early eighteenth century that this view of the mountains started changing even in the common perception. What is often mentioned as a first sign of this change is the journey across the Alps made by the Englishman John Dennis in 1686. For the first time, an author here speaks of "delightful Horror" and "terrible Joy" (Dennis 1693, 33). The feelings of fear and bewilderment caused by a threatening landscape are no longer exclusively presented in negative terms, but are also regarded as a source of pleasure, albeit of a different sort than the pleasure caused by beauty. The feeling of the sublime in those years passed from the rhetorical domain, to which it had been confined for 2,000 years, into the broader aesthetic sphere, becoming a central element of seventeenth-century poetics. Albrecht Haller's 1732 poem on the Alps marked the consecration of the new outlook on wilderness, paving the way for countless literary variations as well as – at a later stage – a new pictorial *vogue*. This was given full expression and widely promoted by Jean-Jacques Rousseau, who in the novel *Julie, or the New Heloise*, sang the praises of high mountains and their moral influence on man (Rousseau 1761, letter 23).

Alongside the sublime, a new aesthetic category emerged in the eighteenth century as a way of marking a break from what had hitherto been understood as "beautiful" nature – that is a nature that is well-arranged, chiefly for cultivation. The new category was that of the *picturesque*, a term which originally meant "suited to making a fine subject in painting". In particular, it referred to rough, jagged, dark landscapes, in contrast to the smooth, regular and sunlit countryside.

One example of picturesque art is first of all provided by Salvator Rosa's *vedute*, in which a varied and irregular nature, often filled with forests, crags and caverns – a fine shelter for brigands and other villains – provides a new paradigm for the landscape. Unlike the sublime, which Kant (1790: § 28) described on the one hand as the boundlessness of nature – unreachable mountains and ocean expanses – and, on the other, as nature's power – storms, volcanoes and floods – the picturesque more modestly emphasizes irregular nature, a rugged, jagged land as opposed to an orderly, flat or only slightly sloping landscape. A round and gently sloping hill or a flowery meadow will be regarded as beautiful; a moor dotted with clusters of trees and streaked with gorges and ravines will be perceived as picturesque. The cultivated countryside, then, might still be considered beautiful, but neither picturesque nor sublime.

A neat exemplification of this change of taste may be found in the history of the garden. While the architectural, geometric, well-ordered garden to some extent represents an extension of the cultivated countryside, and vice-versa, the Mannerist garden – exemplified in Italy by the Pratolino gardens and even more by those of Bomarzo – identifies a "third" nature, that is a mixture of wild and cultivated nature, in which the artificial aspect of the whole is clearly visible.

However, the most decisive break with the paradigm of beautiful cultivated nature was surely made by the picturesque garden and the English garden. Significantly known as "the landscape garden", this was designed in such a way as to conceal its underlying artificiality and to create the impression of pure, wild nature. The gardens surrounding villas and castles, or the country mansions of English aristocrats, were not conceived as agricultural estates – unlike French and Italian gardens, which in a way stood for an intensification or magnification of agricultural processes – but were rather intended to be perceived, as far as possible, as a disorderly and spontaneous nature.

The landscape garden anticipated by a few decades the Romantic garden, which was to establish as mainstream the predilection for wild, rugged and dark nature – as represented in the arts by a new love for mountain vistas (with the paintings of Alexander Cozens in the eighteenth century), frozen landscapes (as in some of Caspar David Friedrich's paintings) and of stormy seas (as in Turner's seascapes). What we find here is no longer the serene nature favored by the Classical landscape painting of Poussin, Lorrain or indeed – well into the eighteenth century – Hackert; we rather find a violent, inhospitable nature. It is no longer a pleasant and charming landscape in which one would like to live, but a barren, stark or threatening landscape.

The idea of conceiving the actual landscape as a projection of landscape painting started spreading precisely in the early nineteenth century and completed the process whereby the "beautiful" landscape had gradually come to be separated from the agricultural one. The gap thus created between the kind of landscape to be admired, painted and described, and cultivated farmland was destined to remain open for almost two centuries. In fact, judging from the works of some contemporary environmental artists, like Richard Long or Denis Oppenheim, fond of hiking and dizzying heights, we might say that the gap remains open to this day.

The fall of love for agricultural landscapes and the rise of the romantic passion for wilderness

There are many reasons why agricultural landscapes have ceased to be considered "beautiful". First of all, what contributes to the disrepute of the agrarian landscape is the still widely held assumption that the only landscapes of genuine aesthetic worth are "extraordinary" landscapes – uncommon, rare and exceptional ones. This tendency obviously runs against the perception of the agricultural landscape as an aesthetically pleasing one, since by definition it is a well-arranged landscape, shaped by everyday practices. If only landscapes of outstanding beauty are regarded as worthy of consideration, then what will be privileged will invariably be landscapes foreign to common human transactions, landscapes that we can only find by moving away not just from the city but also from the countryside – for example by climbing the great heights of mountains or venturing into dangerous areas.

A second reason is probably to be sought in the endurance of an opposition entrenched in common perception: the opposition between the useful and the beautiful. Although everyday experience teaches us that the two values, usefulness and beauty, do not necessarily stand in mutual contrast, and that an object, such as a building, may very well serve a specific function while at the same time constituting an artwork, with regard to the landscape the prejudiced assumption is still that only a landscape serving no utilitarian end can be beautiful – a landscaped not designed for human well-being, an unproductive one.

A third reason, which in a way is the counterpart of the second one, emerges from the observation that usually people who live and work within a given landscape, exploiting it for their own purposes, have no eye for its beauty. One might recall here Cézanne's observations on Mount Sainte-Victoire: Cézanne portrayed it countless times, with boundless love and devotion, on each occasion seeking to delve a little further into his beloved landscape. Yet when speaking with local farmers, he found it impossible to elicit the faintest hint of wonder or admiration for the mountain from them. That was the place of their everyday labor, not a magnificent setting. Farmers, at any rate traditional farmers, typically did not appreciate the "landscape" as such. Indeed, the landscape was usually only "discovered" and valued by burghers who spent their leisure time in the countryside, or by nobles who chose to leave their city palaces for their country mansions. The love of the landscape went hand in hand with the spread of an urban culture: paradoxically, it was city living that nourished the love of the countryside.

In the case of the European landscape (and the Italian one in particular) what has partially balanced the loss of repute of the agrarian landscape, even in the past, has been the acute awareness of the historical and cultural character of the landscape, and hence some appreciation of the role played by agricultural labor with respect to its transformation and conformation. But elsewhere these scruples were missing. Let us think, for instance, of the extent to which the national conscience of the United States has been shaped by the myth of the wilderness, and by the identification of the national spirit with the *natural* and *wild* roots of the environment in which it developed. While the protection of nature emerged in Europe as the protection of natural beauty, in North America it took the form of the conservation of the pristine environment, of nature yet untouched by human labor. The first large natural parks were established in America in the second half of the nineteenth century: *nature*, in a way, replaced *history* as a communal bond. Hence, it represented a nature utterly *different* from history – not the kind of nature that encompasses human labor but the kind that rules it out or, at any rate, makes it impossible on account of its own boundless might and vastness. This is the nature of the big parks of Yellowstone and Yosemite, for example.

Even landscape laws have long borne witness to this marginalization of the cultivated landscape. Consider the case of Italy, where a pervasive and indissoluble link exists between landscape and agriculture and yet the protection of the landscape has long revolved around the idea of *natural* landscape rather than that of an extraordinary combination of natural elements and artificial, historical ones. The undoubtedly significant Italian law 29 June 1939, n. 1497, on the protection of natural beauties, still had *picturesque* beauty as its point of reference, since it explicitly referred to "panoramic beauties regarded as paintings" (Law 29 June 1939, n. 1497, art. 1). Clearly, as one would expect, this law was still based on an acknowledgement of *exceptional* beauty, since it focused its conservation efforts on "fixed features that possess *conspicuous* qualities of natural beauty or geological uniqueness" (Law 29 June 1939, n. 1497, art. 1).

Even the far more recent, and equally praiseworthy, law 8 August 1985, n.431, in defence of areas of relevant environmental interest, operates within a context in which no trace of the agrarian landscape is apparently to be found. This law protects the coastline and the shores of inland waters, particularly "mountains above 1600 meters in the Alps and above 1200 meters in the Appennines", along with "glaciers, parks, forests, volcanoes and wetlands" (Law August 1985, n. 431, art. 1). One might say that conservation begins where agriculture ends.

Recent trends in the appreciation of nature

Over the last 25 years things have taken a somewhat different turn. Farmland is no longer perceived as something opposed to the landscape from an aesthetic perspective. Beauty is no longer exclusively sought in areas where we can harbor the illusion that no visible traces are left by mankind. Of course, I am not referring to the awareness by landscape theorists that European landscape is a *cultural* landscape and hence a *cultivated* one – such awareness has always been widespread. What I am referring to is a newly common appreciation of the countryside, including farmland, as a landscape.

We can easily identify some of the reasons behind this change too. We immediately come across two reasons that may seem irreconcilable at first but are in fact compatible. The first may be described as the relinquishing of the privilege formerly assigned to *exceptional* landscapes. Not just current theories but also current views of the landscape increasingly tend to assign value even to ordinary landscapes, not only to places of exceptional beauty. What is increasingly taking root is the belief that the landscape can be described as an interconnected network, as opposed to the sporadic emergence of beauties as extraordinary as they are mutually unrelated.

A typical example of this new way of perceiving the landscape is to be found in the assumptions made by the *European Landscape Convention* (ELC) of 2000. The ELC tends to consider the landscape as being coextensive with the local territory, incorporating both agricultural landscape and wilderness. The Convention, moreover, explicitly acknowledges that any stretch of a given territory carries an aesthetic identity, thereby acknowledging the existence not just of excellent landscapes but also of common or degraded ones.

This is something that we experience in our everyday lives as well. Most of us agree that a landscape conveys an aesthetic experience not just when we are elated at the sight of landscapes of outstanding beauty and harmony, but also when we are saddened at the sight of spoiled, disfigured and desolate landscapes in which we would never want to live.

By acknowledging the landscape as an essential component of peoples' living environment, the ELC liberates the agrarian landscape from its minority status, just as the Italian *Codice dei beni culturali e del paesaggio* does by identifying the landscape as a "territory that expresses an identity" (D.L. 22 January 2004, n.42, art. 131). The presence of different degrees of value within the landscape is reflected by the multiplicity of possible courses of action identified by the ELC: from the *protection* of landscapes of exceptional significance and beauty to the *management* of common landscapes to the *planning* of degraded ones (ELC, art. 2 and 3).

The second reason for the rise in appreciation of the agrarian landscape, which apparently stands in contrast to the one just illustrated, is the fact that farmland has now become a *rare asset*. In developed countries there is less and less farmland: the number of cultivated plots of land is constantly dwindling. The case of Italy is particularly instructive. Utilized Agricultural Area (UAA) in Italy has decreased by 17.6 percent between 1990 and 2005 (Settis 2010, 8). Contrary to what people often believe or write, this drop is not only due to over-development, that is, to the construction of new houses, roads, shopping malls, sports centers, and the like. In fact, in quantitative terms the main factor accounting for the shrinking of UAA in Italy is the extension of woodland, which has increased considerably in recent decades. From an environmentalist perspective, this might seem like a positive development; yet it is worth bearing in mind that these lands are often unmanaged and thus exposed to fires, poor water control, the spreading of non-native plants turned weeds, and other environmentally undesirable phenomena.

Ultimately, the dwindling of agricultural land in Italy is not so much due to over-development but rather to the depopulation of the countryside and the abandonment of marginal areas, especially mountain ones, which turn into un unmanaged woodlands. After the peak in cultivated land reached in the mid-twentieth century, the number of agricultural plots of land has steadily decreased.

These data concerning farmland should further be combined with those pertaining to the number of agricultural workers, which is also progressively diminishing. The number of people working in the agricultural sector dropped from 4.9 percent in 1999 to 3.6 percent in 2013 (ISTAT 2014). In 1950, agricultural laborers still accounted for 30 percent of the overall workforce.

The cultural consequences of this decline are not always adequately taken into account. Whereas two generations ago most families still had a close connection to the countryside (for instance, by having a father or mother with a rural background), today almost the whole of the Italian population (but this is true in many other developed countries) has no direct connection with the world of farming, which has therefore become elusive for most people. As a consequence, many now perceive the cultivated countryside as a new and unusual environment worth discovering. Perceptual factors too contribute to this new assimilation of the agricultural landscape to the unproductive one conventionally associated with aesthetic experiences. Silence and solitude, which are defining features of standard views of the beautiful and congenial landscape, by now are also associated with the cultivated countryside – at any rate, with the extensive one in which the agricultural labor is concentrated in a few days per hectare, with a small number of farmhands.

Yet another reason for the renewed appreciation of the agrarian landscape has to do with environmental concerns. Agriculture increasingly appears to be, to many Italians at least, a crucial way of safeguarding the landscape. No matter how widespread the mistrust towards agriculture that entails the use of chemicals, it still remains true that agriculture, in all its forms, is *the only artificial use of soil that is also reversible*. Agricultural land is always available land, whereas developed land is lost forever (on a human timescale) unless expensive and complex land reclamation procedures are adopted.

This points to an interesting feature of the Italian landscape, a feature that is however not unique to Italy and should thus stimulate reflection in many other countries as well. Precisely because the Italian landscape is almost entirely shaped by the relation between agricultural labor and nature, agriculture is crucial to conservation. This is shown by the spread of woodlands that I mentioned above: this will form a "natural" landscape where humans have no hand, but that landscape may be extremely *unnatural* for Italy, as it lends its territory a configuration that is utterly alien to its traditional layout. Generally speaking, within the world of agriculture an increasing awareness of this responsibility has emerged, along with methods of cultivation compatible with the local environment and landscape. Environmentalists, too, are beginning to appreciate agriculture as a means to conservation, at least when it comes to landscapes.

The next reason for the renewed appreciation of agrarian landscapes is the demise of a once more clear-cut contrast between city and countryside, urban dwelling and country home. Here we refer not to the phenomenon of urban sprawl, which rather leads to a degraded, "third" type of landscape (to employ the term introduced by Gilles Clément), but to the increase in residential mobility and new forms of rural habitation whereby a considerable percentage of city dwellers choose the countryside as their fixed or frequent abode.

Alongside the new perception of the countryside displayed by outsiders who choose it as their place of residence essentially for its aesthetic qualities and wholesomeness, we are also observing a more marked emphasis being placed on immaterial values in agricultural economic activities, such as those connected to landscape inhabitation, transformation, and management, One example is the growing phenomenon of agri-tourism, where the attractiveness of the landscape clearly plays a prominent role. But let us also think of the emphasis on environmental and landscape qualities that comes with many typical food products, as a way of lending them a unique "aura". By now, even EU policies are taking into account the environmental and landscape function of agriculture (as opposed to its exclusively environmental one), by promoting traditional methods of cultivation, cross-compliance and "greening practices" (Regulation (EU) 1307/2013, art. 43–47).

The *European Rural Heritage Observation Guide*, for example, explicitly associates the value of the landscape with the preservation of agricultural environments, including not only the countryside and methods of cultivation, but also rural buildings and artefacts connected to these activities (CEMAT 2003). The emergence of a new sensitivity is further reflected by the fact that many recently established parks are not merely "environmental" parks located in uncultivated areas, but also include agricultural areas. Italian examples are Parco delle Cinque Terre in Liguria and the Parco del Ticino between Piedmont and Lombardy.

A recent book and two films can provide a confirmation of what we have said. The book is Giorgio Boatti's *Un paese ben coltivato* (Boatti 2014), which explores several Italian regions to identify the new kind of farmer, far from indifferent to the landscape and its safeguarding, whom I have referred to as a new rural dweller. The two films, both of the same year 2014, are centered on country life: Alice Rohrwacher's *The Wonders* and Jonathan Nossiter's *Natural Resistance*. In these films, the directors successfully combine an interest in particular settings with a focus on two typical agricultural productions, possibly the most ancient ones within our civilisation alongside oil production – honey and wine. These two tales, associating the most deep-rooted rural traditions in Italy with new, unexpected protagonists, support the main claim of this paper: namely, that a new sensibility for landscape is becoming more and more widespread in Europe, expanding the appreciation not only of wild nature but also of cultivated land.

References

Baldo, G. and Cazzuffi, E. (eds.). 2013. *Regionis Forma Pulcherrima. Percezioni, lessico, categorie del paesaggio nella letteratura latina*. Florence: Olschki.

Berque, A. 1995. *Les raisons du paysage: de la Chine antique aux environnements de synthèse*. Paris: Hazan.

Boatti, G. 2014. *Un paese ben coltivato. Viaggio nell'Italia che torna alla terra e, forse, a se stessa*. Roma-Bari: Laterza.

Bodei, R. 2005. *Paesaggi sublimi. Gli uomini davanti alla natura selvaggia*. Milan: Bompiani.

CEMAT 2003. *European rural heritage observation guide*. Council of Europe.

De Brosses, C. 1858. *Le Président de Brosses en Italie. Lettres familières écrites d'Italie en 1739 et 1740*. Paris: Didier.

Dennis, J. 1693. *Miscellanies in verse and prose*. London: Knapton.

Di Bene, A. and D'Eusebio, L. (eds.). 2007. *Paesaggio agrario: una questione non risolta*. Rome: Gangemi.

ELC. 2000. *European landscape convention*. Florence, October 20, 2000.

Ferriolo, M., V. 2002. *Etiche del paesaggio. Il progetto del mondo umano.* Roma: editori Riuniti.

ISTAT. 2014. Italia in cifre. Accessed October 10, 2017 from www.istat.it/it/files/2014/10/ItaliaInCifre2014.pdf.

Kant, I. 1963 (1790). *Kritik der Urteilskraft.* Stuttgart: Reclam.

Regulation (EU) No. 1307/2013 of the European Parliament and of the Council. December 17, 2013.

Ritter, J. 1963. *Landschaft. Zur Funktion des Aesthetischen in der modernen Gesellschaft.* Münster: Aschendorff.

Rousseau, J.J. 1978 (1761). *Julie oder die neue Héloise. Briefe zweier Liebenden aus einer kleinen Stadt am Fuße der Alpen.* München: Winkler.

Sereni, E. 1961. *Storia del paesaggio agrario italiano.* Bari: Laterza.

Settis, S. 2010. *Paesaggio Costituzione Cemento.* Turin: Einaudi.

12 Resonance with nature and its loss

Angelika Krebs

Introduction

This paper explores the *aesthetic case* for landscape conservation. The main claim is that the experience of beautiful landscapes is an essential part of human flourishing; it is not just an enriching option for all of us and, certainly, not merely a subjective preference for some of us. Beautiful landscapes can make us feel at home in the world; this constitutes their great and irreplaceable value. The aesthetic argument does not have to carry all the burden of justifying our conservationist attitude. Its role is limited but important. The argument provides a metaphysically innocent understanding of our feeling that we are part of nature, and should try to fit in rather than stand out of it. The argument also serves to underpin, in a deeply humanistic fashion, our horror at the ever-growing grey crust that threatens to cover all of the Earth's surface.[1]

As a first step, I will clarify the *concept of landscape*: a landscape is a larger natural whole whose principle of unity is its *gestalt*, character, *stimmung*, mood, or atmosphere. Which plants grow in a landscape and how they look depending on season is essential for the atmosphere of the landscape. I will then distinguish various ways in which we experience landscape atmosphere, namely perception, empathy, sympathy, and infection. I do this to lay the ground for the specifically aesthetic claim that I put forward in the last and main section: that when we experience the atmosphere of a landscape *aesthetically*, we respond to it by resonance: by feeling at home.

Before I embark on the rocky road ahead, let me get you into the right mood by presenting a passage from German writer Peter Kurzeck's autobiographical 2003 novel *Als Gast* ("As a guest"). In the passage, the author recalls a walk he took with a friend in the city forest of Frankfurt, which – apart from the highway and its hum in the background – is as empty and quiet "as if the earth [. . .] had long stopped speaking to us." Kurzeck's written language is like music. It is "deep-acting." It touches you immediately. This is why it can give us "knowledge by acquaintance" and make us *feel* the loss of nature:

> Through the piece of forest now, along its edge. Such a scanty little forest – however one walks, one always walks along the edge. And the forest as if emptied out. Rather as if just erected, you say to yourself. No roots? Without

> roots, the trees? Professionally put up by professionals. Quality forest. Guaranteed to last. Life-size. And secured with care. Like the real thing. Exactly almost like the real thing! And so quiet, as if the earth, every spot of earth, the plants, the stones and every thing, as if all the world had long stopped speaking to us. And then we had stopped too. A long time ago. We do not answer! So quiet, but behind the quiet a hum, a growing hum. From all sides. And coming towards us. Or as inside one's own head.
>
> (Kurzeck 2012, 191–92, translation by A. Krebs)

Kurzeck's walk does not lead through a forest, but only through a piece of forest. A real forest is large; you can enter deep inside it. Such an inside does not exist in a piece of forest. A piece of forest is not a forest anymore.

Among the trees in the piece of forest, there is nothing left: no undergrowth, no shrubs, no mushrooms. Even the trees no longer look real – they seem fake, highly praised in the excited language of advertising that culminates in the paradoxical cry: "Exactly almost like the real thing!"

We are unable to resonate with such trees, with such a piece of forest, with such highly artificial and cut up nature. It seems that in a world like this, we also have stopped resonating with and between ourselves. Yet, behind this dead quiet, the cars on the motorway are roaring. The machine world seems to be the only thing that still grows, like nature once did. The machine threatens us. It intoxicates us.

The concept of landscape

To clarify the concept of landscape we must first look at the concept of nature. As Aristotle already taught us in his *Physics*, *nature* is that part of the world which has not been made by human beings but comes into existence and vanishes by itself. Artifacts are the opposites of nature in this sense: they are made by human beings. The distinction between nature and artifacts is polar, or gradual (like the distinction between bright and dark) and not binary, or dichotomous (like the dis tinction between being pregnant and not pregnant): one cannot be a little bit pregnant, but one can be in a more or less dark or bright place. There is hardly any untouched nature left on Earth today. Most of what we call nature, the conservation of which we are concerned with, lies, in fact, between the extremes of pure nature and pure artifact. It is a mix of natural and artificial, in which the first aspect still prevails.

In nature, we can distinguish natural organisms and things (like plants and rocks) from larger natural units (like landscapes, parks or gardens). There is no sharp boundary between landscapes and gardens. *Gardens* are, first, laid out for aesthetic enjoyment and in this respect they fall somewhere between art and nature; they also usually surround a house, and are themselves surrounded by a fence, so that they mediate between the house and the landscape.

Landscapes that, unlike gardens, are relatively free from human ends are especially pertinent to the experience of natural resonance. Yet, one can certainly also resonate with nature in gardens and parks, as well as with singular organisms and natural things. But our focus here is on landscapes.

In twelfth century Old High German, "lantscaf" denoted a large natural area and its population. In the fifteenth-century Netherlands, the term could also refer to a painting of a large natural unit. Art historians still talk of landscapes in these terms. Today, the boundaries of landscapes are no longer just physical (and thus political), as they often were in the past. Following a suggestion that the German philosopher and sociologist Georg Simmel made in his classic piece on the philosophy of landscape a hundred years ago, for us today it is *atmosphere* or *character* that constitutes the *unity* of landscapes (Simmel 2007).

Where a large natural area loses its character because of a natural catastrophe or human destruction, it lacks the unity necessary for being a landscape. It turns into an expressionless *heterogeneity*, into a non-place or into landscape waste. It does not turn into an ugly (that is, aesthetically unattractive, not beautiful) landscape.

Not every landscape change amounts to landscape destruction, either. The change can also be for the better: the Golden Gate Bridge in San Francisco might be an example of a positive landscape change. Still, much of what is going on around us today does amount to landscape destruction.[2]

Perception, empathy, sympathy, and infection

In order to prepare the ground for the specifically *aesthetic* experience of landscape, four basic types of landscape experience should be distinguished: perception (or understanding), empathy (or vicarious/reproduced feeling), sympathy (or fellow feeling), and infection.[3]

When we *perceive* that a landscape is peaceful, we remain affectively more or less neutral. We simply realize that it is peaceful.

When we *empathize* with a peaceful landscape, we move with its atmosphere, enacting it but not sharing it. Empathy occupies an intermediate position between perception and sympathy. Cruel people are not sympathetic to the suffering of their victims, but they still need empathy in order to fully enjoy their victims' pain.

When we *sympathize* with a peaceful landscape, we move with its atmosphere and share it. We resonate emotionally, as we do when listening to a favorite piece of music. Sympathy is an emotion in the full sense, including bodily feeling, cognitive evaluation and behaviour; empathy, n the other hand, is "only" a vivid mode of cognitive understanding.[4] Sympathy comes in two variants: participatory sympathy and meta-sympathy. Only the first is relevant for landscapes. When meta-sympathetic, we are sad (for example) *about* the sadness or the bad predicament of another, but we do not accompany her through her own sadness, as in the first participatory variant.[5]

When we are *infected* by a peaceful landscape, we are swayed by its atmosphere. Infection is causal, while perception, empathy and sympathy are intentional; they are directed towards the expressive quality of the landscape. In being directed towards an "other," they presuppose some distance between self and other. Infection, on the other hand, annuls this distance. Infection is highly relevant for mental health and wellness, but it does not play much of a role in aesthetic experience.

Feeling at home in nature

This final section spells out how resonating aesthetically with landscape atmospheres can make us feel at home in the world. It distinguishes between stronger and weaker understandings of feeling at home. While beauty, especially functional beauty, allows for feeling perfectly at home, sublimity affords only a partial or ambivalent version of it.

Aesthetic landscape experience involves not only turning one's attention to a landscape, perceiving its atmosphere and empathizing with it, but also sympathetically sharing its atmosphere *for its own sake*. In stressing the "intrinsicness" of aesthetic experience, as well as the distance that is constituent of perception, empathy and sympathy (as they are directed towards an "other"), this understanding is reminiscent of Kant's aesthetics – though Kant's aesthetics is much colder than that. As John Dewey, among many others, observed, sympathetic emotions play no role in Kant: "To define the emotional element of esthetic perception merely as the pleasure taken in the act of contemplation, independent of what is excited by the matter contemplated, results, however, in a thoroughly anaemic conception of art" (Dewey 2005, 264). Instead of aesthetic contemplation, I therefore prefer to speak of aesthetic "resonance" (I elaborate below). Still, it is important to distinguish, with Kant, between aesthetic experience on the one hand and physiological and psychological impact, or effect, on the other. The main thesis of this paper about how aesthetically attractive landscapes can make us feel at home in the world does not concern *causal impact* or *effect*. Rather, it concerns the quality of the aesthetic experience itself, which can include, as a by-product, the mood of feeling at home.

Like all intrinsic (in the sense just specified) activities, aesthetic sympathetic attention or resonance is accompanied by pleasure – a disinterested kind of pleasure, or in Kant's terms, "interesseloses Wohlgefallen." Georg Henrik von Wright calls this kind of pleasure "active pleasure" and contrasts it, first, with "passive pleasure," such as the good taste of an apple, and, second, with the "pleasure of satisfaction," that is the feeling we have when we achieve what we want (cf. Wright 1993, 63–65). Similarly, Roger Scruton calls active pleasure "pleasure in," and contrasts it with "pleasure from" and "pleasure that" (cf. Scruton 2009, 28–31). Active pleasure or pleasure in is something we cannot intentionally induce: it arises only when we are absorbed in an activity and forget about our daily worries. It is a *by-product*.

Mihaly Csikszentmihalyi (1990) has popularized the notion of the self-forgetful drive, which characterizes active pleasure as *flow*. Csikszentmihalyi has also presented empirical findings to show how, in some particularly successful cases of actively pleasant intrinsic activities, subjects become aware of themselves as parts of a larger whole. According to Csikszentmihalyi, there is nothing esoteric or metaphysical in that:

> When a person invests all her psychic energy into an interaction – whether it is with another person, a boat, a mountain, or a piece of music – she in effect becomes part of a system of action greater than what the individual self had been before.
>
> (Csikszentmihalyi 1990, 65)

Martin Buber in *I and Thou* (2004) describes the same phenomenon in different terms, when he writes about the mystic:

> What the ecstatic man calls union is the enrapturing dynamic of relation, not a unity arisen in this moment of the world's time that dissolves the *I* and *Thou*, but the dynamic of relation itself, which can put itself before its bearers as they steadily confront one another, and cover each from the feeling of the other enraptured one. Here, then, on the brink, the relational act goes beyond itself; the relation itself in its vital unity is felt so forcibly that its parts seem to fade before it, and in the force of its life, the *I* and the *Thou*, between which it is established, are forgotten.
>
> (Buber 2004, 69)

Another example of this phenomenon familiar to musicians is the feeling you can get when, after having rehearsed a symphony with the orchestra, it suddenly seems as if the music or the orchestra *plays you*. Before, it was you who played your instrument together with all the others who played theirs. Now, you feel as being part of the whole. And this is a new experience, exceeding the active pleasure you have felt all along in rehearsing the symphony with the other musicians.

Thus, the affective quality of aesthetic experience highlighted so far lies in sympathy and in flow, plus, in some cases, the feeling of being at home. The *metaphor of resonance* underlines this affective quality.

However, the metaphor of resonance might be misleading in at least three ways. First, physical resonance occurs when one object vibrates with another at the same or a similar natural frequency, for example when the G- and D-strings of a violin vibrate with a G-major chord on a piano. This is a *causal* phenomenon, whereas aesthetic resonance is first and foremost intentional sympathy.

Second, physical resonance is not only causal but also *instantaneous*; aesthetic resonance, in contrast, requires a "gymnastics of attention" (to borrow a phrase from Roger Scruton). It takes time and effort; only sometimes, in learnt spontaneity, does it occur instantaneously. We can distinguish the immediate seizure produced by an aesthetic atmosphere from the discrimination that sets in afterwards, which may or may not validate the first, immediate impression. This first impression is still directed, however, and is not to be confused with infection.

Third, physical resonance tends to be *bilateral* (and even amplifying: think of the famous example of marching soldiers collapsing a bridge). Not only does the violin resonate with the piano, the piano resonates back with the violin. This has led Hartmut Rosa to conceptualize resonance in general, that is any kind of resonance – be it causal (physiological), biographical, social, relational (as in love and friendship), or aesthetic – as a mutual phenomenon. For Rosa resonance is a response relation, which requires both parties to speak with their own voices. In aesthetic resonance, not only do we respond to the world, the world also responds to us (cf. Rosa 2016, 298). This mutual concept of aesthetic resonance, however, slips into metaphysics, as nature does not respond to us in any literal sense. Aesthetic resonance with nature is a one-way connection. What Rosa might have in mind is the Eichendorffian phenomenon of the magic word, which sounds the song that is sleeping in all things. Soberly understood, this phenomenon is nothing but our feeling that our metaphors *fit* the world. We create our metaphors but we cannot create the fit. The fit must happen by itself. If it does, it feels as if the world responds to us, and begins to sing.

Beauty

Landscapes are *beautiful*, in a broad sense, when they invite and reward intrinsic sympathetic attention or resonance. Their appeal is not limited to some of us but open to all. Aesthetic landscape resonance is not just a subjective preference: it is a universally (and perhaps particularly easily) accessible form that the desire for beauty can assume. The desire for beauty is an anthropological constant: man does not live on bread alone. Fulfilling this desire in one way or another is an important part of the good human life. As morality requires respect for the essentials of the good life of all human beings, conserving beauty is a moral obligation.

How beautiful landscapes and other beautiful objects or ensembles manage to lure and satisfy us is, of course, the central question of aesthetics. Classical answers stress symmetry, harmony, or unity in diversity. Modern answers focus on the experiencing subject. On a Kantian account, beautiful objects or ensembles bring our faculties of imagination, "Einbildungskraft," and understanding, "Verstand," into free play. This overly intellectual model should at least be complemented by the idea of a "free play of sympathy," to borrow again a phrase from Roger Scruton (cf. Scruton 1997, 355). It is not only our cognitive faculties that are attracted and challenged by beauty, but also our affective powers. Beauty does not only make us think about many things, it also makes us feel many things. It makes us open up and grow both rationally and emotionally.

Do the atmospheric and the beautiful then amount to the same thing (at least with regard to beautiful landscapes and nonrepresentational forms of art)? Not necessarily. Something might have a strong positive or negative atmosphere in the sense of an overwhelming impact, infecting us but not inviting us to attend to it and sympathize with it for its own sake. We might also formulate this point differently, as follows: what is merely atmospheric *has* an atmosphere, while what is beautiful *expresses* an atmosphere. Putting it like this, however, might lead us to weaken the notion of "expression," allowing us to say that beautiful landscapes express atmospheres, to the point of obscuring a difference between landscapes and works of art that seems important. Artworks are expressive in a different and deeper sense than landscapes are. Expressive art is a kind of communication. It pursues meaning. It articulates, explores and meditates on human concepts in a structure of its own. Expressiveness in art is an achievement. This does not true for landscapes. Compared with art, the expressiveness of landscapes is a superficial phenomenon.

Still, landscape beauty is special and *cannot be replaced* by other kinds of beauty. If it were replaceable, nothing much would follow from the aesthetic argument in terms of landscape conservation. One reason why landscape beauty is special is that we experience landscapes synaesthetically, that is feel them with *all our senses*, not only with our eyes and ears, which are more capable of aesthetic distance than our noses, tongues and fingers are. We even move around in landscapes. Hiking is a popular way to appreciate natural beauty. Sensual feeling and, yes, infection is part and parcel of aesthetic landscape experience (see Jonas 1954). We can thus add infection to the affective aspects of aesthetic landscape experience outlined so far, which include sympathy, flow, and feeling at home. Infection serves to increase the immersive effect of beautiful landscapes, so that we may feel at home *in* them, both sensually and aesthetically.

The other and more striking reason why beautiful landscapes are irreplaceable is that they fulfil our conscious or unconscious longing for being part of and not being alienated from the natural world – the world that is just there, that comes into being and vanishes by itself. No fake of nature can fulfil this longing. We need beautiful landscapes to heal the rift between subject and nature: both nature out there and the nature within us. *Living in harmony with nature* in this sense is more than an enriching option for a good life: it is an essential part of human flourishing. As Martin Heidegger and his follower Otto Friedrich Bollnow put it, beautiful landscapes teach us how to "*dwell on earth*" (cf. Heidegger 2013, 141–60; Bollnow 1988, 44–62). They give us a sense of place and make us honor it. They invite us to put down roots somewhere and identify and care for it as our special "Heimat."

Sublimity

There are stronger and weaker forms of feeling at home in nature. So far, I have mainly talked about the strongest one, *perfect sympathetic coordination*, which feels like unity.

Often, however, we succeed only *partially* in our attempt at sympathetically moving with something. Our failure need not be due to us; it could also be due to the landscape. The classical distinction between the beautiful and the sublime is relevant here. For our purposes, it can be reconstructed as follows. Only the beautiful (now in a more limited sense than before and no longer synonymous with "aesthetically attractive") allows us to be fully taken up in it. The sublime, in its infinite extent and power, entices us to sympathetically move with it, too: the subject enjoys participating in its magnitude and strength. However, the subject also feels painfully reminded of her own insignificance and vulnerability. The sublime confronts us with a tension between a celebration of the landscape and a powerful sense of self-diminution. Still, insofar as the sublime appeals to us and invites us to partially move with it, neither leaving us cold nor threatening us existentially, it is possible to talk about feeling at home, in a weaker sense, in sublime nature too (cf. Cochrane 2012).[6]

Functional beauty

A third experience of feeling at home in nature occurs when we attend to the landscape that surrounds us not as such, but in relation to ourselves, that is, in its functionality for our own good. In Kantian terminology, the latter kind of experience is directed at the "dependent" beauty of the landscape and not at its "pure" beauty. A landscape that looks as if it affords a good human life is beautiful in the functional or corresponsive sense (cf. Seel 1991). It is ugly if it doesn't. Thus, contrary to so-called *positive aesthetics*, there is a sense in which nature can be *ugly* (for positive aesthetics see Budd 2002; Carlson 2000).

In functionally beautiful landscapes we feel at home, not only because they have a good physiological and psychological impact on us, but also because they indicate, by the way they look, sound, and smell, that they can support human life and provide for its needs. *Evolutionary aesthetics*, which traces our sense for beauty back to our sense for landscapes with a high survival value for our species, like the savannah, finds a limited justification here (see Dutton 2009).

Conclusion

To conclude, we are natural beings who long to live in harmony with nature. The on-going destruction of beautiful landscapes turns us into aliens in our own world. Our own world is rapidly becoming so empty and quiet as if the earth had long stopped speaking to us. And then we had stopped too.

Notes

1. For a critical taxonomy of the many arguments for the conservation of nature: cf. Krebs (1999).
2. For a scientific study that documents landscape destruction see Ewald and Klaus (2009). For an artistic exemplification, look at Müller's set of pictures *Alle Jahre wieder saust der Presslufthammer nieder* 1973, or listen to the passages 5 and 8 on the second CD of Peter Kurzeck's audio book *Ein Sommer, der bleibt* 2007.
3. The contemporary debate on empathy, in which "empathy" can mean any of these different phenomena, still needs to regain the conceptual standard that phenomenology reached at the beginning of the last century, most notably in the writings of Max Scheler and Edith Stein (see Scheler 1954; Stein 1989; for a summary and elaboration of Scheler's position, cf. Krebs 2011).

4. Although in certain cases empathy can lead to actual emotions, it does not necessarily do so (cf. Goldie 2000).
5. For this distinction between two different kinds of sympathy cf. the first chapter of Edith Stein's *On the Problem of Empathy* (Stein 1989).
6. As the example of the Alps famously shows, only *existential* threat and sublimity exclude each other.

References

Bollnow, O. F. 1988. Die Stadt, das Grün und der Mensch. Pp. 44–62 in *Zwischen Philosophie und Pädagogik*, edited by Otto Friedrich Bollnow. Aachen: O.F. Weitz.
Buber, M. 2004. *I and thou*. London: Continuum.
Budd, M. 2002. *The aesthetic appreciation of nature*. Oxford: Oxford University Press.
Carlson, A. 2000. *Aesthetics and the environment*. London: Routledge.
Cochrane, T. 2012. The emotional experience of the sublime. *Canadian Journal of Philosophy* 42, no. 2:125–48.
Csikszentmihalyi, M. 1990. *Flow*. New York: Harper & Row.
Dewey, J. 2005. *Art as experience*. London: Perigree.
Dutton, D. 2009. *The art instinct: Beauty, pleasure, and human evolution*. Oxford: Oxford University Press.
Ewald, K. and Klaus, G. 2009. *Die ausgewechselte Landschaft: Vom Umgang der Schweiz mit ihrer wichtigsten natürlichen Ressource*. Bern: Haupt.
Goldie, P. 2000. *The emotions*. Oxford: Clarendon Press.
Heidegger, M. (ed.). 2013. Building dwelling thinking. Pp. 141–60 in *Poetry, language, thought*, edited by M. Heidegger. New York: Harper & Row.
Jonas, H. 1954. The nobility of sight. *Philosophy and Phenomenological Research* 14, no. 4 (June): 207–519.
Krebs, A. 1999. *Ethics of nature: A map*. Berlin: De Gruyter.
Krebs, A. 2011. The phenomenology of shared feeling. *Appraisal* 8, no. 3: 35–50.
Kurzeck, P. 2007. *Ein Sommer, der bleibt*. Berlin: Suppose (audio book).
Kurzeck, P. 2012. *Als Gast*. Frankfurt: Fischer.
Müller, J. 1973. *Alle Jahre wieder saust der Presslufthammer nieder*. Aarau: Sauerländer.
Rosa, H. 2016. *Resonanz*. Frankfurt: Suhrkamp.
Scheler, M. 1954. *The nature of sympathy*. New Haven: Yale University Press.
Scruton, R. 1997. *The aesthetics of music*. Oxford: Oxford University Press.
Scruton, R. 2009. *Beauty*. Oxford: Oxford University Press.
Seel, M. 1991. *Eine Ästhetik der Natur*. Frankfurt: Suhrkamp.
Simmel, G. 2007. The philosophy of landscape. *Theory, Culture and Society* 24, no. 7–8: 20–29.
Stein, E.1989. *On the problem of empathy*. Washington, DC: ICS Publications.
von Wright, G. H. 1993. *The varieties of goodness*. Bristol: Thoemmes Press.

13 Plant risks

Can risk assessment accommodate "cultural services"?

Paul B. Thompson

This chapter is about plant-based risk. It considers whether the conceptual framework of environmental risk assessment can be made sensitive to some hazards that have not previously been well accommodated. Ideally, the chapter would include a more detailed discussion of plant-oriented values that would be positioned within larger debates in environmental philosophy. While I am positively disposed toward theories of value that include botanical organisms as moral subjects, this chapter will not engage with this debate directly. The ethical questions discussed in this chapter are limited to the ways that plants and plant communities (including but not limited to cultivated ecosystems) have significance for human beings. *That* plants have significance for many people will be taken as obvious: gardens, arboretums, natural preserves and also purely decorative use of flowers and greenery are well loved, and this does not even begin to discuss the value that plants attain as the primary basis of human nutriment and (hence) survival. Space limitations in the current format do not permit a more detailed treatment.

Plant ethics: The risk based view

Risks from plant-based sources provides are, as a matter of fact, widely recognized to exist. Their assessment by toxicologists and control through various governmental regulatory authorities demonstrates the existence of generally recognized procedures for conceptualizing the ethical significance of plants, and for taking measures to conserve those values deemed to be most significant. Since plants are not generally taken to be moral subjects, the risk framework derives ethical significance from the way that sentient creatures (and especially humans) encounter plants. Although plants can be a source of nourishment, pleasure and meaning, we more typically identify ethical issues in our use of plants when they pose some chance of causing harm. Many plants contain chemicals that induce toxic effects ranging from nausea, swelling, and rashes to paralysis. Humans have recognized and even utilized plant-derived toxins for millennia, but the potential for inadvertent exposure remains an obvious source of plant-based risk. In addition to toxic chemicals produced within plant tissues, plants can support the growth of toxic microorganisms (though animal products are far more potent source of exposure to bacterial hazards). Without pretension of completeness, a search of the World Wide Web for the terms "plant-based risk" yields the following five categories.

Food safety: All industrial societies have developed institutions for limiting the chance of exposure to toxins through consumption of food. These institutions combine classificatory systems that are based on tradition, rather than science, with elaborate laboratory testing and clinical studies to determine the level of risk associated with foods and food additives. The foundation for food safety risk assessment is the United States Food and Drug Administration list of foods Generally Recognized As Safe (GRAS). Not surprisingly the GRAS list contains all of the plants that we typically associate with food, even though some contain chemicals that are known to cause illness at high doses.

Poison control: Humans and other animals can suffer exposure to plant-based toxins through physical contact or through inadvertent consumption. Poisons are also administered maliciously. Unlike food safety systems that address plant-based risks by attempting to either minimize the chance of exposure or to limit the dose of a plant-based toxin to a safe level, the risks of plant poisoning are address by medical and public health authorities through education and through post-exposure medical treatment.

Veterinary health: Non-humans can, of course, suffer ill effects in the two above categories as well. Plant-based risks to animals of toxic exposure through animal feed or through other types of inadvertent contact are regulated with considerably less vigor than they are for humans. Nevertheless, modern societies recognize these hazards as significant and have developed institutions for coping with them.

Environmental impact: Protecting biodiversity is widely recognized as a legitimate value, and some species of plants enjoy legal protection when they are either threatened with extinction or become especially emblematic of environmental values. Contemporary risk-based regulatory policies protect both plant and animal communities when human activity threatens the survival of an entire species or when a species' presence within a given ecosystem is critical for its stability and continuous functioning.

Genetic modification: There is a widely (though far from universally) held belief that modification of a plant's genome using the suite of techniques that manipulate DNA constitutes a fifth category of plant-based risk.

There are, to be sure, additional harms and hazards that human beings endure as a result of their encounter with plants. Farmers or nursery operators can experience severe economic hardship when plants under their care undergo an unexpected turn for the worse, for example. It is philosophically significant that such outcomes do not surface rapidly in an electronic search of terms records for plant-based risk. One possible explanation for this is that the five categories listed above are all in some sense biological. Perhaps we do not think of outcomes that occur in the economic sphere as owing to our relationship with plants, even when plants and plant health are a proximate cause of them. Perhaps there are strong associations between the word "risk" and biological or health-oriented mechanisms that lead to an exclusion of economic hardships, but if this is so it is striking to note that historical treatments of risk and risk management are strongly tilted toward the fields of insurance and financial investment. Although the social imagination veers toward the biological when the words "plant" and "risk" are conjoined, a brief review of the risk assessment approach provides a basis for a philosophical critique of this tendency.

Risk assessment and social choice

For present purposes *social choice* can be defined as the application of decision theoretic approaches (including ethical theories) to problems where society as a whole can be understood as the choice-making agent. Social choice can be a purely theoretical exercise conducted by academics, or it can be an applied practice in real-world decisions. The social choice perspective is most obviously useful in policymaking contexts where a governmental decision is expected to transcend partisan interests, but it can also be used to comparatively evaluate future scenarios that have not been generated by alternative policy options. Alternative scenarios generated by climate models are sometimes evaluated in terms of their overall risks and benefits, for example (Stern, 2008). Although social choice is philosophically complex (see Sen, 2011), the focus of this section is to clarify how value judgments are deployed in risk assessments that have been developed for application in classic cases of governmental policymaking.

The risk assessment framework can be summarized by the following schematic (see Figure 13.1). *Hazard identification* is the process of identifying what bad outcomes can occur as a result of adopting one option rather than another. In classic regulatory decision making, the option under analysis is often whether to permit the use or release of a new product, or whether to adopt rules governing some technical practice (such as generation of electricity using nuclear reactors of a given design). Hazard identification is a largely inductive process of ascertaining the possibility of an adverse outcome based on the current state of the relevant science, and estimating its severity. Simply knowing what can go wrong does not provide an estimate of risk, however. Hence *risk quantification* utilizes a variety of modeling, statistical and experimental techniques to estimate the probability that a hazard will materialize, given adoption of the option under consideration. Though conceptually simple, both activities can be difficult to accomplish in practice. Some of these difficulties are philosophical and will be taken up below.

Risk analysts will stress a subtle point at this juncture. Risk should not be conflated simply with the loss or "badness" of the hazard, should it occur. Risk must also reflect the chance or probability that hazard actually *will* materialize, and this can only be appreciated once the various factors that contribute to a

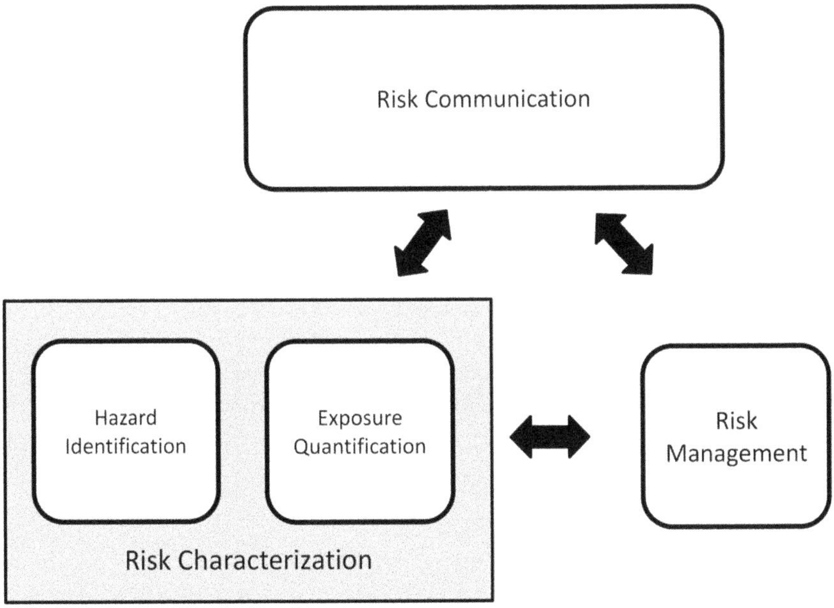

Figure 13.1 The risk assessment schema (Paul B. Thompson)

given individual's or group's exposure to the hazard have been taken into account in the risk quantification phase. Research in the social psychology of decision making suggests that many people tend to neglect or poorly understand the probabilistic dimension of decision-making (Sunstein, 2002), hence there is an argument for discounting the public's estimate of relative risks associated with any given activity or phenomenon. This argument has been made with respect to plant risks especially in connection with genetic modification (Hettlich and Walther, 2012).

Having an estimate of risk does not tell a regulator what to do about it. One option is always "Nothing." Simply accept the risk. Other options include regulatory measures that mitigate severity of hazards, that compensate losers or that develop mechanisms to secure consent from those parties that will bear risks. The procedure for making a decision about risks is represented by the *risk management* element in the schema. *Cost–benefit analysis* is a procedure developed by economists to allow decision makers the opportunity to include not only risks (as determined by hazard identification and risk quantification) but also beneficial elements associated with adopting the option. The key point to notice is that one can make a distinction between *assessing* (or estimating) what the risk is, and deciding what to do about it. The decision making phase is referred to as *risk management*.

Social choice models for decision making under risk that propagated in the 1970s and 1980s tended stop with hazard identification, risk quantification and risk management. These models were criticized as lacking a role for public input. In response, *risk communication* was added to the model following an influential U.S. National Research Council study of the 1990s colloquially referred to as "the orange book" (NRC, 1996). The role of risk communication within the overall framework of risk assessment and social choice continues to be a topic of confusion, contestation and continuing debate. Some participants in those debates would see the question raised by the title of this chapter (e.g. Can risk assessment accommodate cultural services?) might argue that the cultural values (including those associated with plants) should be brought into decision making through the mechanism of risk communication (e.g. Hartley, 2012; Wickson, 2007). The position that will be argued below is that cultural services can be better integrated into the risk assessment logic model as a component of hazard identification. A careful discussion of plant risks and risk communication is beyond the scope of the chapter, though a few suggestive comments will be made in the closing section.

The ontology of plant hazards

Although risk assessment is commonly thought of as a science-based procedure (and I will not dispute this), hazard identification is inherently normative. While a number of biological mechanisms for causing harm are reflected in the dominant view, they are only hazards in the sense that have been judged to be both biologically possible and unwanted, harmful or contrary to the anthropocentric values that frame social choice. There are two points to notice here. First, a possible course of events does not get classified as a hazard unless it is deemed to be "bad," unwanted or in some way, adverse, (Cranor, 1990), Risk is thus a phenomenon that illustrates the deep interpenetration of values into science (Douglas, 2000). Second, the scientific nature of hazard identification in normatively non-controversial outcomes such as poisoning or biodiversity loss derives from the judgment that the outcome in question can be classified as a biologically meaningful outcome. This is, it must be stressed, an *ontological* judgment. The "social imaginary" that frames present-day thinking is categorically structured in a manner that allows for events of many different kinds. Losing a bet on a coin flip is a meaningful event and one that is sometimes the focus of risk assessment. But it is not an event that can be characterized easily in biological terms. The event-space for risk assessment is thus parsed by judgments of ontological classification that, in the exemplary case, distinguish economic and biological orders of being. While philosophers have devoted considerable effort to the normativity of hazard identification (Elliott and Resnik, 2014; Hansson, 2010; Heyman, 2010) there has been considerably less attention given to an ontological judgment that is equally important.

Judgements of ontological classification interpenetrate both hazard identification and the quantification of exposure in risk assessment. Losing a bet is just not the *kind* of bad outcome that qualifies as biologically meaningful. The ontological classification is, in this case, being made at the stage of hazard identification. But other cases involve classification of the mechanisms that determine exposure to events such as mortality and morbidity that are clearly biological in kind. One of the key case studies in risk science involved an exposure to caesium-137 after a medical device was removed from an abandoned clinic in Goiânia, Brazil in 1987. Caesuim-137 is highly toxic and there were several fatalities as a result of the incident, plus an extremely expensive clean-up operation that remains somewhat controversial to this day (Roper and Sperb Leite, 2001). The incident was followed by a panic throughout Brazil due to the erroneous assumption that the fatalities were due to some form of contagion. Although the radioactive contamination was confined to a single neighborhood, people canceled vacations and business trips when the story broke in national newspapers. Airline pilots refused to allow people from Goiânia to board their planes and cars with Goiás license plates were stoned in other parts of Brazil (Simons, 1987). Although radiation poisoning is a very serious hazard, it is not subject to contagion in the manner of an infectious disease. The fears that led to significant social and economic consequences in Goiânia were based on a perception of the hazard that was normatively valid (caesium-137 poisoning is quite serious) but the exposure mechanism (contagion) was judged not biologically possible.

The judgment that hazard and exposure are biologically possible will be deeply informed by the state of scientific knowledge at any given time. One dramatic example is provided by tomatoes and potatoes. The toxicity of Solanaceae plants in the deadly nightshade family is now known be associated with tropane alkaloids. These alkaloids are produced in some plant tissues, but not in others. The biological possibility (e.g. risk) of introducing toxicity through plant breeding is tied to the potential for altering regulatory genes that govern where tropane alkaloids are produced in the process of plant transformation through interbreeding with related varieties (Oksman-Caldentey and Arroo, 2000). Although tomatoes are now enjoyed virtually everywhere, they were avoided for decades based on their relationship to belladonna and some infortuitous acidic interaction with pewter used in European dinnerware (Smith, 2013). Better science has improved our understanding of the risks associated with plants in the nightshade facility, but it is also important to maintain the distinction between the ontological possibility of a hazard and its measured likelihood. Toxic tomatoes are biologically possible, but this is not to say that they are likely.

However, none of this should be taken to suggest that biological possibility is the sole criterion for judging the ontological status of a hazard. The mechanisms for exposure to radiation poisoning are not primarily biological (though the etiology of the hazard is). More pertinent to the topic of this chapter, there are a host of hazards that are grounded in socioeconomic events. Some of them must be classified as plant-based risks. As markets for organic and non-GMO crops have developed, crops intended for these markets can be contaminated by wind-borne pollen drifting from neighboring fields. Farmers who may have incurred significant expenses to produce organic and non-GMO crops can lose a significant amount of the crop's economic value when contamination occurs. This potential for economic loss is, from the farmer's perspective, unarguably an important risk that will have significant weight in decision making about whether to participate in these markets or not (Bruce, 2003; Ervin, Glenna and Jussaume, 2010).

To put this point in a more abstract and somewhat more philosophical manner, ontological possibility must be judged according to the mechanisms appropriate to the ontological status of the risk in question. Biological possibility must be ascertained in reference to metabolic and ecological pathways. These will depend upon but are not strictly reducible to biophysical and chemical reactions. Socioeconomic hazards will depend upon markets and exchange relations that are (similarly) not reducible to physics and chemistry. Put succinctly, hazard identification as currently practiced requires a pluralistic ontology that rejects strong reductionism. Reductive ontologies that interpret existence predicates solely in terms of physics and chemistry are simply not plausible for the purposes of hazard identification.

It is possible, however, to go further in calling for ontological pluralism. There are many features of our social world that reside primarily in our discursive practices. In other contexts, I have argued that people are quite rational and entirely justified in feeling that they are "at risk" when dealing with others who are judged to be power-seeking or who have defective moral character. Taking a walk or being stranded in a "bad neighborhood" might be judged risky because one imagines a straightforward biophysical hazard such as bodily assault, but the vulnerability one feels in such situations is not simply a matter of judging that such an outcome is more likely in one place, rather than another. In a similar way, people feel "at risk" when they lose confidence in actors controlling food production. They see their situation as fraught in significant part *because* these actors are not judged to be trustworthy, even when they are quite unspecific as to what hazards might ensue (Thompson, 1997; 2012). Risks to cultural values might be of this sort, but they might also derive from mechanisms that reduce the frequency of certain cultural food practices, or that undercut the social context in which food practices occur. The general point is that given an appropriately pluralistic ontology, there is no reason why these discursive outcomes cannot be included among the hazards included in the hazard identification stage of a risk assessment.

Ontology and policy

Deploying the cultural "we" implied by a shared social imagination, we seem to have accepted the need for governmental oversight of biological risks, while denying (or at least questioning) the legitimacy of regulating economic risks. The social institutionalization of these ontological category judgements is hard to trace but one can catch sight of it in John Stuart Mill's essay *On Liberty*. After laying out the utilitarian rationale for his principle of non-interference in liberty of conscience, Mill explains how use of the police power to constrain actions that cause harm to others is fully justified. But Mill excludes a class of harms from his overall argument. The losses we suffer as a result of fair competition simply do not count morally (Mill, 1859). This characterization requires quite a bit of unpacking, of course, and a significant portion of the political debate that has ensued since Mill published his essay in the middle of the nineteenth century has been devoted to it. Governments have, in fact, attempted to regulate the exemplary risk of losing a coin flip (or at least they have regulated commercial gambling in many societies), and important elements of the disagreement between capitalism and socialism concern the role of government in modulating certain economic risks.

The considerable complexity of this large sociopolitical controversy can be at least partially constrained by focusing on the topic of this chapter, the assessment and regulation of plant risks. The biological hazards already discussed are widely regarded as appropriate targets for regulation, though accompanying judgments about the relatively low probability of hazards materializing through the biological mechanisms of using GRAS plants or plant products limit the need for formal risk assessment and policy judgment in a large number of cases. Debates over hazard and exposure in connection with biological risks from transgenic or cysgenic modification, gene editing or gene drives would be on-going subjects of controversy in the implementation of these generally accepted principles. At the same time, there are well recognized economic hazards associated with plant production, and here there are also significant public expenditures devoted to assessing the nature and extent of these risks. Policymakers have eschewed attempts to shield economic agents entirely from the risks of economic loss in connection with plant production, but there are extensive (and complex) policies that moderate these risks. These policies range from crop insurance programs to elaborate subsidy payments triggered by fluctuation in the commodity price for agricultural products.

The economists who have estimated and managed the financial risks of plant production may not have made explicit use of the risk assessment rubric, but their approach is nonetheless consistent with it. Hazard identification has involved recognizing various types of economic loss. These include the failure to recoup monies expended on seed, fertilizer and other inputs, as well as the potential for a decline in the asset value of farmland or farming equipment. Economic models are intended to assign probability distributions to these financial hazards, producing a characterization of economic risk that is fully consonant with the hazard plus exposure model described by the rubric (see Figure 13.1). There are then both policy and individual risk management decisions to be made, ranging from simply taking one's chances to moderating risks by doing things that reduce the exposure to a particular hazard or compensating losers through subsidy or insurance payments. Much of the farm extension work that has been done by economists for almost a century would qualify as a prime example of risk communication.

Thus, if the dominant view is that only biological hazards constitute legitimate plant risks, this view is not, in fact, consistent with actual social practice. In practice, industrial societies have assessed and managed risk in according to an ontology that recognizes events in the biophysical health-oriented order and also events in the financial or economic order as potential hazards. Toxicological hazards, animal health and environmental impacts are recognized plant risks, to be sure, but so are crop loss and fluctuations in the value of plant-production assets. These ontologically distinct risks are managed differently, as well they should be. It is as if our practice exhibits greater ontological fluency than our speech. But if economic hazards can be accommodated under the rubric of risk assessment, what are we to say about the role that eating plants or plant products plays in establishing basic cultural forms?

The plant-oriented use values discussed in the first section of the chapter exist in an ontological order that is neither biological nor economic. It is a socially constituted order of being that would benefit from considerably more exposition than I have been able to undertake in this chapter. It is a kind of value that we designate as broadly "cultural". Cultural value derives from the way that social habits and shared meanings serve as an enabling platform for symbolic interactions of a diverse kind, including, one might add, the economic transactions that give rise to the seemingly more obvious forms of economic risk that have just been discussed. Cultural values can be placed "at risk" as readily as those values more conventionally examined in risk assessment, though we should probably admit that we are much less sure about the mechanisms that expose us to the potential loss of cultural heritage and communicative capacity than we are for biological and economic hazards.

The recently evolved notion of ecosystem services provides one way that we might hypothesize exposure to the hazard of cultural identity loss. We might, in other words, conceptualize the cultural identity formation provided by our shared use of certain plants or plant products as food or feed as a service that our plant communities have been providing to us. We could then speculate on what would be involved if we had to replace this service that plants have done for us, were they to be altered or changed in ways that these meaning giving experiences no longer occur as a matter of course. Plants are already being conceptualized in this manner in so far as they function as components of place or landscape (Tengberg et al. 2012). While I would maintain some skepticism about the adequacy of existing economic models for adequate measurement of these costs, the framework of cultural services would provide an opening for philosophical discussion of the way in which plants contribute to human well-being in deeply underappreciated ways.

It is also quite unclear how the cultural services we derive from plants might be reflected in a policy setting. The goal of this chapter has been to sketch and begin the argument for an approach that is broadly compatible with risk-oriented ways of thinking about the moral significance of plants. Contrary to the suggestion that we have no precedent for plant ethics, risk assessment has accorded explicit significance to plant-based hazards in both biological and economic orders of being. If there has been confusion or obscurity with respect to these hazards, it is due largely to a practiced tendency to ignore or downplay the implicit ontological commitments of hazard identification as it is currently practiced by toxicologists, food scientists, ecologists and economists. Once these commitments are made explicit, it becomes possible to recognize their normative significance and to undertake a more conventionally ethical discussion about how to handle them through alternative policy means. Once the buried ontological commitments are uncovered, in other words, plant ethics can proceed under the rubric of risk management and risk communication.

Conclusion

In summary, it is entirely possible to consider the "cultural services" that people derive from plants within the overarching framework of risk assessment. This answer to the title question of this chapter has been developed by first characterizing a cultural service in terms of the aesthetic enjoyment people derive from both cultivated and uncultivated plants, as well as the culinary traditions associated with the production and consumption of icon and culturally meaningful foods. Given this starting point, it is, in fact, important to recognize that mainstream (or dominant) views of plant-based risk emphasize biologically based hazards such as toxicity or disruption of an ecosystem's functional integrity. This is the current "social imaginary" that structures the conduct of plant scientists and government regulators alike. It is a construction of nature that provides little place for considering even socioeconomic outcomes, let alone cultural services.

However, a review of the logic model on which risk assessment is based begins to open the door for an alternative way to think about cultural services. Risk assessment is conceptualized as comprising four stages: hazard identification, risk quantification, risk management and risk communication. Hazard identification is a key stage in that phenomena not judged as both possible and detrimental will not be recognized as hazards. While philosophers and social scientists have given significant attention to the values-basis for judging an outcome to be a hazard, this chapter has argued that exclusion of cultural services is more likely to occur based on the way that ontological criteria for possibility are developed and applied. Succinctly stated, in our practice we seem to have become biophysical reductionists, effectively presuming that if an outcome cannot be described and its likelihood assessed through the analysis of biological and physiochemical mechanisms, it simply does not exist.

This is patently absurd, and it is doubtful that any biologist or risk analyst would defend a practice so baldly characterized. I have indeed attempted to give a more nuanced and careful development of this theme in the body of the chapter, and a still more thorough treatment might explore how the ontology of hazard identification is constrained less by the decision making of biological experts than by the autocratic dictates of risk managers. As noted above, risk management is where the decision about what one is going to do about a risk is made, but there is little point in collecting significant information about a class of hazards if the risk managers have already decided that they are not going include any of that information in their decision making. A significant amount of the critical work on risk policy has called for better risk communication, meaning more participation by affected parties in the process of assessing risks, as well as in deciding what to do about them (Ben-Ari and Or-Chen, 2009;

Wickson, 2007). The perspective developed in this chapter has suggested that better hazard identification might actually suffice to ameliorate some of the important issues. This should not, however, be interpreted as an argument against risk communication. It is rather to say that engagement with stakeholders and non-expert members of the public will be most felicitous when it is recognized that they may well be concerned about hazards that have *not* been taken into consideration, even if they might be.

One final conclusion can be stated even more broadly. A new consideration of plant ethics need not and should not be conceived as antithetical to the basic logic model of risk analysis. The model has flexibilities that could be and should be more widely exploited than they typically are. Cultural change and the loss of cultural values that are tied to the plants we eat and our recreational enjoyment of plant communities are fully capable of being studied within the model offered by hazard identification, even if the science for modeling the mechanisms of such change remains underdeveloped. We are indeed constrained by our current "social imaginary." There are, in that sense, ways in which our conceptualizations of risk are constrained by our collective confidence in thinking that an organism without a central nervous system cannot be a moral subject. My reflections in this paper have not challenged that presumption, but the questions of ontological possibility only become more crucial (and more fascinating) when we move far beyond the limits of our present-day moral imagination. I have not intended to say anything that would hamper such a move.

References

Ben-Ari, A., and Or-Chen, K. 2009. Integrating competing conceptions of risk: A call for future direction of research. *Journal of Risk Research* 12: 865–77.

Bruce, D. 2003. Contamination, crop trials, and compatibility. *Journal of Agricultural and Environmental Ethics* 16: 595–604.

Cranor, C. 1990. Moral issues in risk assessment, *Ethics* 101: 123–43.

Douglas, H. 2000. Inductive risk and values in science. *Philosophy of Science*, 67: 559–79.

Elliott, K. C. and Resnik, D. B. 2014. Science, policy, and the transparency of values. *Environmental Health Perspectives* 122: 647–50.

Ervin, D. E., Glenna, L. L. and Jussaume, R. A. 2010. Are biotechnology and sustainable agriculture compatible? *Renewable Agriculture and Food Systems* 25: 143–57.

Hansson, S. O., 2010. Risk: objective or subjective, facts or values. *Journal of Risk Research* 13: 231–8.

Hartley, J. 2012. *Communication, cultural and media studies: The key concepts*. New York: Routledge.

Hettich, P. and Walther, S. 2012. Bleak prospects for research in GMP in Switzerland. *European Journal of Risk Regulation* 3(3): 361–71.

Heyman, B. 2010. Values and health risks. In *Risk, safety, and clinical practice: Health care through the lens of risk*, edited by B. Heyman, M. Shaw, A. Alaszewski and M. Titterton. New York: Oxford University Press.

Mill, J. S. 1859. *On liberty*. London: John W. Parker and Son.

NRC (National Research Council). 1996. *Understanding risk: Informing decisions in a democratic society*. Washington DC: National Academy Press.

Roper, L. D. and Sperb Leite, M. A. 2001. The Goiânia radiation incident: A failure of science and society. *Veneer Magazine*. Accessed July 3, 2017 from www.veneermagazine.com/01-18/01/the_group/goiania.html.

Sen, A. 2011. *The idea of justice*. Cambridge, MA: Harvard University Press.

Simons, M. 1987. Radiation fears infect Brazil after accident. *The New York Times*. Accessed July 3, 2017 from www.nytimes.com/1987/12/02/world/radiation-fears-infect-brazil-after-accident.html?mcubz=2.

Stern, N. 2008. The stern review on the economics of climate change. Accessed July 9, 2017 from http://webarchive.nationalarchives.gov.uk/20080910140413/http://www.hm-treasury.gov.uk/independent_reviews/stern_review_economics_climate_change/sternreview_index.cfm.

Sunstein, C. 2002. *Risk and reason: Safety, law and the environment*. New York: Cambridge University Press.

Tengberg, A., Fredholm, S., Eliasson, I., Knez, I., Saltzman, K. and Wetterberg, O. 2012. Cultural ecosystem services provided by landscapes: assessment of heritage values and identity. *Ecosystem Services* 2: 14–26.

Thompson, P. B. 1997. Science policy and moral purity: The case of animal biotechnology. *Agriculture and Human Values* 14: 11–27.

Thompson, P. B. 2007. *Food biotechnology in ethical perspective*, 2nd ed. Dordrecht, NL: Springer.

Thompson, P. B. 2012. Synthetic biology needs a synthetic bioethics. *Ethics, Policy and Environment* 15: 1–20.

Wickson, F. 2007. From risk to uncertainty in the regulation of GMOs: Social theory and Australian practice. *New Genetics and Society*, 26: 325–39.

14 Utopia in the garden

New utopian and dystopian thinking in current debates on nature, agriculture, and food

Christian Dürnberger

Telling stories of an alternative and the end

Current times are characterized by particularly weighty controversies concerning the relationship between man and nature. Debates on topics such as food, agriculture, climate change and environmental degradation not only raise the question of individual measures but also pose the principal question of the possibility of an alternative society and a different way of life. Debates about technology, which is central both as a cause and a possible remedy of the ecological challenges that we currently face, are often not just concerned with specific issues but also articulate the larger question of what kind of world we want to live in. In particular, technological innovations with the potential to impact the distant future trigger reflections on the scenarios that are being painted by current trends. The desirability of such scenarios is far from ensured. Because precise prognoses regarding the long-term effects of current trends are not possible, however, and since the question of the possibility of a different way of life always requires a narrative answer (cf. Schwonke 1957, 105)[1], societal discourses recur to a traditional, basic cultural tool to make sense of such complex, seemingly abstract questions: recounting visions of utopia and dystopia.

In this article, I trace utopian – from the Greek "ou" (no) and "topos" (place) – and dystopian – from the Greek "dys" (bad) "topos" (place) – content in current disputes over nature, agriculture and food. Narratives of "the good alternative" as much as narratives of looming disaster are both parts of the rhetoric of these disputes. I begin with a historical overview of utopian and dystopian thinking, and go on to situate such thinking within current disputes over nature, agriculture and food. I suggest that a "new" utopian thinking can be identified which does not tell a story about the ideal form of government or a new societal order, as utopia often did in the past: it rather tells a story of utopia at the small scale of individual life – a utopia *en miniature*.

Utopia and its alleged end

The term "utopia" was coined by Thomas More in the year 1516 (cf. More 2016). As mentioned, "Utopia" means literally "no-place". In common parlance, "utopia" typically refers to some vision of imagined community or society with highly desirable qualities. Utopias play with conceivable, alternative forms of social existence. Against the background of the actual conditions, they design an alternative model of a better (or even perfect) society, making deficits and aberrations of the present even more apparent. Starting from criticism of the existing, utopian thinking creates parallel worlds that do not yet exist but are in principal possible and, at least from their authors' point of view, highly desirable.

Such utopian concepts can be identified through the whole history of ideas. Following a broad understanding of the term, narrations of paradises lost, such as in the Bible's *Genesis* or Ovid's *Metamorphoses*, have utopian content.[2] Another early landmark of utopian thinking is Plato's *Republic* (approx. 370 B.C.), where an ideal state is described in detail. But utopian thinking undoubtedly reached its prime during the early modern period. Three authors deserve mention above all. First, Thomas More with his powerful narrative of the island *Utopia* (1516) – which, as noted, is the eponym of the whole genre. Second, Tommaso Campanella, who painted a visionary social order without any private property, which he intended as an earthly image of the divine order in *The City of the Sun* (1602). Third, Francis Bacon, who described in *The New Atlantis* (1627) his ideal of a society that followed and realized his lifetime commitment to scientific research and technological progress.

Talking about a flowering period also means talking about a phase of fading: It has been said that utopia has ended. Jürgen Habermas diagnosed an "exhaustion of utopian energies"[3] in the mid-1980s (Habermas 1985, 141). The breakdown of communism – which could be understood as a large-scale attempt to realize a social utopia – allowed thinkers to speak about the definitive end of the utopian age (cf. Fest 1991).

The current place of utopian thinking is therefore more in science fiction literature than it is in social-philosophical works. However, it is not only failed practical tests that drained away the power of utopian visions. Utopias lost their appeal for at least three further reasons:

1. At least since Popper's *The Open Society and its Enemies*, first published 1945, utopian content has been criticized as potentially totalitarian. Popper exemplarily argued that the powerful state machinery depicted in Plato's *Republic*, controlling and regimenting the everyday life of citizens, is distinctive of totalitarian regimes. The priority that these regimes often assign to the collective over the individual undermines liberty and is typically inhumane (cf. Popper 1992).
2. Utopias have also been questioned on grounds of feasibility. The historian Lucian Hölscher (1985, 404ff) has shown how the role of utopia, and the usage of the term, changed over time specifically on feasibility grounds.

Whereas in earlier centuries utopian stories were generously employed, especially in view of their expected pedagogical benefits, utopias were called to respond of their own feasibility from the nineteenth century onwards (cf. Hölscher 1985, 404ff.). "This is utopian" became more and more a way to point out that something is unrealistic. Whether utopias are impracticable by definition (as the etymology suggests) is an open question. Herbert Marcuse recommended reserving the term for projects that are both unrealistic and simply impossible (cf. Marcuse 1980, 11).

Yet utopias always seem to have an in-built aspiration to implementation. On the one hand, utopian ideals can be and often are implemented: Representative examples include Robert Owen's colony *New Harmony*, the communities inspired by the visions of Charles Fourier, or the Icarian movement which tried to put into practice the utopian novel *Travel and Adventures of Lord William Carisdall in Icaria* (1840), by Étienne Cabet (cf. Saage 2002, 345–70). Contemporary attempts can also be found, for example in the *Equilibrismus* movement.[4] On the other hand, even utopias that are never tried out in practice can influence the existing situation by functioning as a kind of "regulative principle" (Saage 2002, 345) or as "concrete utopia" (cf. Bloch 1980). As Jörn Rüsen describes them, utopias are always "constituted by hopes, which go beyond the realms of current answers to what is feasible, without fundamentally questioning the possibility of fulfiling the hope" (Rüsen 1985, 358).

3. Utopia have lost appeal also because they seem to be dramaturgically unappealing, even boring, compared to dystopias. Of course, what is appealing or unappealing from a dramaturgical point of view is a matter of dispute; however, the increasing number of dystopian stories and visions, particularly in the twentieth century, is certainly remarkable. These artistic works on the potential futures of humanity, and of being human, typically choose dystopian tones. Examples abound, from Yevgeny Zamyatin's *We* (1920) to Aldous Huxley's *Brave New World* (1932), from George Orwell's *1984* (1949) to Margaret Atwood's *The Handmaid's Tale* (1985). All these works paint designed social systems that render dramaturgically rich, dark pictures of the future. The drama is increased by the fact that these dystopias all indicate that the sources of the undesirable future developments can already be found in the present.

In sum, utopian sketches of an ideal society and way of life are suspect in terms of democracy, removed from reality in terms of feasibility, and seen as comparatively boring and un-dynamic in dramaturgical terms. Richard Saage, one of the most important scholars of utopias in the recent past, contests the suggestion that utopia has ended, however. He criticizes as unimaginative the narrow understanding that associates utopias with authoritarian, state-dominated universal concepts of an ideal political system. Accordingly, he argues that Popper's criticism only addresses a specific form of utopian tradition, while the wide variety of utopian thinking is ignored (cf. Saage 1995, 76ff). According to Saage, doing without utopian thinking means cultural impoverishment; furthermore, utopias still have orientating potential:

> Admittedly, those who understand the future just as an extension of the present don't need any utopia. However, those who understand it as characterized by open horizons must say what the world we want to live in tomorrow should look like. How should we answer this question in other ways than by a utopian fiction?
>
> (Saage 1995, 99)

Generally, utopias – divested from totalitarian undertones – seem still able to create a forum for discussing existing systems and beliefs as well as possible alternatives at various scales. The narration of alternative (utopian and dystopian) visions to the existing system may be particularly useful today that the trope "there is no alternative" seems prevalent.[5]

Current controversies in light of utopian and dystopian thinking

In the following, I argue that we can improve our understanding of complex controversies about nature, agriculture and food by considering their relations to utopian and dystopian thinking. I will do this in four steps: (1) I propose that conflicts concerning the environment and technologies are (often implicitly) influenced by utopian and dystopian thinking; (2) I focus on the shifts in the role of technology in utopian and dystopian sketches as well as on (3) the relationship between man and nature as a new essential motif in this context; finally (4), I raise the question of whether there is a new form of utopia *en miniature*, especially when it comes to topics such as nature, agriculture and food.

From statistical prognoses to narrations

Already in the late 1950s, Schwonke stated that utopian and dystopian ideas about the future of scientific-technical society cannot only be found in science fiction literature but also in countless socio-political debates relating to energy, technology and everyday life (cf. Schwonke 1957, III). These ideas also occur in places where one would not expect them. To give two examples:

1. Birnbacher and Schicha show the extent to which the dispute about concrete claims in the context of sustainability debates can be interpreted according to two opposing paradigm visions of the future (cf. Birnbacher and Schicha 2001, 19ff.). According to the optimistic paradigm, future generations will be generally better off than current generations because of the general pro-cess of technological, social, and institutional progress. All that present generations must do, therefore, is not frustrating this process. On the pessimistic paradigm, the responsibility of current generations is far greater, as the path humanity has taken is negative and future generations will find

themselves in worsened conditions, due to precipitating ecological and social circumstances, unless present generations do something about it (Birnbacher and Schicha suggest that the pessimistic paradigm is currently comparatively more *en vogue* in sustainability debates).

2. Utopian and dystopian ideas can also be found in the emotional debate over green genetic engineering – a debate that concerns all three domains: nature, agriculture and food. The proponents sketch a utopia of plants that are ideally tailored to our needs and contribute considerably to the world's nutrition, with relevant technologies used in a responsible and well-monitored way in view of social improvement. In contrast, opponents choose dystopian language to present biotechnological methods in food production as steps into a dark future of dependence on large seed companies, "superweeds", the end of small-scale agriculture, the complete transformation of nature into culture, and so on. For example, when Jeremy Rifkin outlines the future of agriculture driven by the new biotechnologies in his popular bestseller *The Biotech Century*, his description may appear dystopian to many readers. According to Rifkin, future agriculture could produce food in huge bacteria tanks (cf. Rifkin 2000, 24). Such an image of agriculture contradicts influential societal conceptions of how food should be produced – namely on a small farm, where farming is an idyllic and contemplative activity without high-tech interventions (cf. Dürnberger 2008, 45–52).

The debate over biotechnology in agriculture is therefore not only about the risks and benefits of currently available methods but is also about different images of the future: "Proponents of biotechnology promise paradise, while the opponents prophesy hell." (Korthals 2003, 354). While trying to convince the other side, or the undecided, statistical prognoses are accompanied and enriched by lively, easy-to-remember stories and visions. Sometimes these visions are hopeful, sometimes they are worrisome. Both sides are telling stories in an attempt to answer the question of where present developments will end up and to convey their answers to the public. Looking at these attempts, the question arises if utopian and dystopian sketches are not just elements of an intentional communication strategy, but maybe also inevitable components of controversies on technologies that cannot rely on precise predictions about their long-term consequences.

From Bacon to Orwell: Perspectives on technology

Science and technology are at the epicenter of late modern and current utopian literature. Future human well-being and happiness are no longer sought in the best form of government and social order (as, for example, in Plato's *Republic*), but rather in technical and scientific progress.[6] Francis Bacon was a pioneer in this context. In his *The New Atlantis* (1627), European seafarers discover an island after surviving a storm. An islander, serving as travel guide, introduces the

European guests to the (ideal) society on the island, which is founded upon the immense success of the local sciences. The so-called *Salomon's House* is not only the center of the scientific community but also the center of the whole society. It is described as "the lantern of this kingdom" (Bacon 2001, 30). The tasks of Salomon's House are congruent with Bacon's lifetime project of transforming science, from contemplative observation to an active understanding of natural processes that may change and influence them. Although *Salomon's House* is "dedicated to the study of the works and creatures of God" (Bacon 2001, 30), its mission also betrays a rather more applied understanding of science: "The end of our foundation is the knowledge of causes, and secret motions of things; and the enlarging of the bounds of human empire, to the effecting of all things possible" (Bacon 2001, 52).

The technological achievements on Bacon's island are indeed impressive. For example, the travel guide presents the successes of plant breeding thus:

> And we make (by art) in the same orchards and gardens, trees and flowers to come earlier or later than their seasons; and to come up and bear more speedily than by their natural course they do. We make them also by art greater much than their nature; and their fruit greater and sweeter and of differing taste, smell, colour, and figure, from their nature. And many of them we so order, as they become of medicinal use. We have also means to make divers plants rise by mixtures of earths without seeds; and likewise to make divers new plants, differing from the vulgar; and to make one tree or plant turn into another.
>
> (Bacon 2001, 61)

It seems that the inhabitants of *New Atlantis* would accept and like the possibilities of green genetic engineering to an extent unimaginable even today.

Since Bacon, technology as a great hope for a better future has been a vital trope in the utopian genre. Dina Brandt showed, for example, the extent to which German utopian novels in the first half of the twentieth century were marked by a strong technological optimism (cf. Brandt 2007). Today, authors cannot display such optimism without being suspected of naïve tech-euphoria.

Later literature has been dark, and created numerous dystopian stories focusing especially on the power of modern technologies to change human bodies and psychology, societies, and the planet in unwanted ways. In Yevgeny Zamyatin's novel *We* (1924), for example, new medical technologies allow for the disablement of the part of the brain presiding over imagination. Aldous Huxley's *Brave New World* (1932) vividly discloses the dystopian potential of reproductive medicine. In George Orwell's *1984*, modern communication technology leads to totalitarian monitoring and control over citizens. In that respect, the genre of dystopian novels can be understood as an expression of the end of technological optimism – characterized by a general loss of trust in the benefits of technologies (cf. Sieferle 1984, 11).[7]

In various controversies that concern the man-nature relationship, there is also a tendency towards dystopian interpretations of specific technologies. This is, for example, in evidence in the debate on biotechnology in agriculture: green genetic engineering is seen as the final step to a complete mechanization of life, a technology that will be out of control in the near or far future, from which only few powerful agents will benefit, and so on (cf. Dürnberger 2012).

From Bogdanov to Callenbach: Perspectives on nature

The current utopian genre plays a motif that is unknown in the classical utopian tradition: ecological reconciliation. Contemporary humans find themselves in an unprecedented situation of global ecological distress. Since the twentieth century, technologies have developed that have the potential to threaten all living beings on this planet (cf. Jonas 1984, 15ff). Accordingly, much of our thinking about nature has come to emphasize its frailty and need for protection (cf. Schäfer 1996, 199–226).

This is visible in current utopian and dystopian narrations. In the classical utopian genre of the early modern period, the ideal society is based essentially on dominance over nature. In fact, Saage emphasizes that one can comprehend the anti-individualistic preference for the collective, which is so important to many utopias of the early modern period, only by keeping in mind that a collective fight against nature is always assumed to be raging in the background (cf. Saage 1995, 239ff). More, Campanella, and Bacon are united in their ambition to win this battle.

An illustrative example of Saage's argument can be found in the work of the Russian philosopher and writer Alexander Bogdanov (1873–1928). In his novel *Red Star* (1908), Bogdanov tells the story of a communist society on Mars. The local relationship between man and nature is described as follows: "We have peace between people, that is true, but there is no peace with elementary nature. Such a peace is not possible. Nature is our enemy, which has to be defeated again and again and again" (Bogdanow 1989, 77).

This perspective has now radically changed. Beginning in the second half of the twentieth century, we have had clear experiences of the fragility of nature, and stories about utopian societies have begun to also involve visions of a sustainable peace between man and nature. In this context, the most influential ecological utopia is the novel *Ecotopia* (1975) by Ernest Callenbach (1929–2012). To summarize its storyline: After the collapse of the previous system, a state named *Ecotopia* is established on the west coast of the former USA. Decades after a self-imposed isolation, the state allows the journalist William Weston to travel through the new country. Weston describes – at first skeptically, later with growing enthusiasm – the local society, which is oriented in all areas of life towards the idea of sustainability. In telling this story, *Ecotopia* does not promote an eco-romantic return to the roots but opens a critical discussion on the consequences of technology for both man and nature (cf. Callenbach 1996).

Ecological reconciliation is also a central motif of utopias that have been tried out in practice. An early example, focusing on food sufficiency, can be found in the German Empire. At the end of the nineteenth century, the so-called "Lebensreformer" (*reformers of life*) movement founded the vegetarian orcharding colony *Eden* outside the gates of Berlin. Its general aim was to create a new form of living away from urban stress and annoyance. More precisely, the movement demanded vegetarianism, healthy and home-grown food, new teaching concepts for schools, and a simple life in and with nature (cf. Radisch 2010, 72–80).

Typically, however, the literature that comments on the man-nature relationship is much darker. Apocalyptic[8] images are part of any culture and epoch, but their presence is particularly strong in current stories about man-made ecological catastrophes.[9] Especially among adherents of nature conservation movements, a change of imagery towards apocalyptical visions can be identified from the early 1970s (cf. Wöbse 2005, 246) and all the way to the present (cf. Brimblecombe 2004, 50). It is unclear what the actual importance and force of such imagery is. As Joachim Radkau (2002, 305) states: "Does the ecological movement really need the fear of the end of the world? The historic experience seems to show that practical ethics does not need the belief in hell."

The already mentioned debate on green genetic engineering is also heavily influenced by dystopian images and visions. Some critics fear the complete transformation of nature into artifact. A nature robbed of its independence and exposed to human arbitrariness is seen as a great loss. When Bernhard Gills argues that a concept of nature as alternative to culture is an ideal surface for the projection of dreams, and especially for dreams about a better society (cf. Gill 2003, 75ff.), one can add: nature is an ideal projection-surface not only for dreams but also for nightmares.

From the ideal state to lifestyle as a utopian space of possibilities

Most of the above discussion speaks the language of the classical utopian tradition. Although new issues such as ecological reconciliation are introduced, the core of utopian thinking remains tied to the tradition of More: by telling a story about an ideal or dreadful fictitious society, present tendencies of the real world are critically discussed. Whenever utopia is said to have ended, its defenders underline the *diversity* of utopian thinking. Taking this diversity seriously also means asking for utopian thinking beyond the classical tradition. In particular, debates on nature, agriculture and food seem to be interesting fields for this effort, since they explicitly or implicitly discuss alternative visions of an ideal man–nature relationship.

These debates have a dimension that is typically missing in the traditional, more socially-focused utopian genre. A utopia about the ideal man–nature relationship does not need to develop a perfect political constitution: the ideal can also be reached via an a-political withdrawal and seclusion from social life. In other words, an "ideal" (e.g. sustainable or respectful) relationship with nature also stands open for the individual. (In fact, chances are that, in the contemporary world, it is even easier to develop such a relationship as individuals than as societies.)

In this context, utopian thinking does not present visions of a social order as alternatives to the existing system: instead, it attempts to develop a "private" utopia. In other words, the alternative is not designed as a draft for a political constitution, but as an attempt to live an alternative lifestyle that is a practical criticism to the current system. One could say: *L'utopie, c'est moi.* The ecological movement is familiar with these kinds of dropouts in favor of small-scaled utopia. One clear example is Henry David Thoreau (1817–1862) – a US writer, philosopher and icon of the conservation movement who cocooned in a self-built log cabin at Walden for 2 years, to get a feeling of what living in and with nature was (cf. Thoreau 2009).

The dropout only represents one instance of a broader movement that shows tendencies towards withdrawal, particularly when it comes to questions of food and nature. These days, numerous so-called self-supporters document on blogs their attempts to eat only what they have grown; books on self-sufficiency are booming; and community-supported agriculture projects try to bridge the gap between farmers and consumers. Furthermore, urban gardening is spreading to small areas in cities, not only because of a new love for allotment gardens but also because of idealistic goals, such as achieving self-sufficiency, promoting intercultural exchange, using urban areas to produce food for low-income families or for pursuing energy-saving and climate-friendly means of food production (cf. Müller 2011). The ambition is often the autarkic, local production of nutrition for a modest number of people.[10]

Such attempts can be read in light of their utopian content. Based on a critique of the existing system, and maybe also tired by the "overdose" of dystopian visions in this field, these experiments want to show a better yet feasible alternative. They do not suggest a new system of government or social order, but rather open a small-scale utopian space of possibilities. This kind of "private" utopia was described as "a-political" above, but this needs to be qualified. Although this new utopian thinking does not deliver large-scale alternative models of an ideal society, it delivers many attempted realizations of "small" individual ideals that still have the potential (and sometimes the explicit aim) of having a political impact. The departure into a new age and a new society remains a vital utopian goal.

Summary and outlook

Concerning conflicts over technology and the environment, it must be kept in mind that societal debates revolve around not only benefits and risks but also utopian and dystopian visions. Controversies about nature, agriculture and food can be understood as posing the question of the possibility of alternative lives and societies. The general debate on nature conservation, and the discussion on biotechnology in plant breeding, are both relevantly characterized by utopian and dystopian narratives about the future of agriculture and the future of the man–nature relationship.

These narratives should be developed and explored in the right places. Utopias and dystopias provide the opportunity to transcend existing systems and current tendencies and to freely express fears, concerns, and hopes.[11] This opportunity is important, particularly when it comes to emotionally charged debates that play with the imagination. Yet narratives should be presented explicitly as narratives.[12] A sharp line must be drawn between telling stories of utopian and dystopian scenarios on the one side, and passing ethical judgments on the other.[13]

Concerning research in utopian genre, the argument that utopian thinking has ended should be rejected. Utopias today disclose new central human issues, above all with respect to ecological questions. Utopian as well as dystopian visions are influencing controversies about new technologies and the environment, but in these debates, and more generally in debates about the man–nature relationship, utopian thinking is currently to be found less in the design of ideal social systems and more in reflections on individual ways of life.

Schwonke has suggested that utopian thinking has shifted from drafts of political constitutions to science fiction literature (cf. Schwonke 1957, 42ff.). Yet science fiction is not the only place where utopian thinking can be found today. In fact, debates and questions of lifestyle regarding nature, agriculture and food seem to be an ideal biotope for a kind of thinking which can be described as utopias *en miniature*. This means that the much-cited "exhaustion of utopian energies" is an exhaustion of the traditional social utopia. Utopian thinking itself is not exhausted at all; it only withdrew from public places – into the garden.

Acknowledgements

The original version of this article was first published in German in 2014 in "Projektion Natur. Grüne Gentechnik im Fokus der Wissenschaften" (edited by Annette Meyer and Stephan Schleissing) at Vandenhoeck and Ruprecht, Göttingen. It was written within the context of the Bavarian research association "ForPlanta", funded by the Bavarian Ministry for Science, Research and Arts. I want to thank Stephan Schleissing (Institute TTN at the LMU Munich) as well as the project team "New Directions in Plant Ethics" (Angela Kallhoff, Marcello Di Paola, Maria Schörgenhumer and Wolfgang Damoser) for their important feedback and support.

Notes

1. Questions concerning the "whole" life or a good way of living are challenging philosophical thinking, but also go beyond and affect narratives; even scientific facts and extrapolations often need pictorial and vivid representations and images to reach societies.
2. The utopian content of the Bible is most explicit in the chiliastic hope for paradise's return at the end of time (cf. Seibt 1985, 273).
3. All quotes in the article are translated by the author, except the quotations of Bacon.
4. Equilibrismus is a movement with its own socio-ecological economic model. The movement follows three phases, from concept through fiction to reality. After laying

out a theoretical foundation (cf. Freystedt and Bihl 2009) and a literal description of the vision (cf. Fleck 2009), the movement is planning to put its utopian vision into practice on a small island in the South Seas for at least 10 years (www.equilibrismus.de).
5. In addition, Rüsen underlines the cathartic function of utopias when he notes that they are like dreams humans must have in order to deal with the world and with themselves (cf. Rüsen 1985, 359).
6. According to Schwonke, this new focus, driven by the natural sciences and technology, is leading to a change of the whole genre. Utopian thinking is shifting from designs of political constitutions to science fiction literature (cf. Schwonke 1957, 42ff).
7. To what extent an unconditional technological optimism ever existed is debatable. Having said this, historical analyses are able to show that past discourses tended toward a much more positive judgment of technological transformation of nature than is usually the case today (cf. exemplarily Blackbourn 2008).
8. At this point, my usage of the term "apocalypse" follows everyday language, which understands it as a synonym for the end of the world. In theological contexts, the term includes the momentum of a new beginning and redemption.
9. The collapse of civilization as a consequence of the exploitation of nature is a popular topos in literature, as is the motif of nature's revenge. For example, in the science fiction classic *The Day Of The Triffids* (1951), by John Wyndham, highly cultivated plants take revenge against humans and start to kill them.
10. Following early Foucault, the question arises to what extent such projects can also be understood as *heterotopias*. Foucault defines heterotopias as real and effective places that are part of the society and at the same time something like an alternative draft of the society (cf. Foucault 1993, 39).
11. In addition, current ethical debates underline the power and effects of narrations for moral reflections (cf. Ricoeur and Greisch 1996, 201).
12. Of course, this appeal calls for some kind of operationalization. An appropriate institutionalization and consideration of utopian and dystopian concepts and images in societal and political discussions is therefore anything but a trivial enterprise.
13. In this context, the so-called *slippery slope* argument is a good example. This argument alleges that a number of minor decisions in the present will lead to a morally unacceptable future situation. From an ethical perspective, there are two queries: Do the minor decisions necessarily lead to the sketched scenario, or does the claim rather reveal a lack of confidence in political regulation? Second, is the future scenario indeed morally unacceptable, and if so, for what reasons?

References

Bacon, F. 2001. *The new Atlantis*. New York: P.F. Collier & Son.

Birnbacher, D. and Schicha, C. 2001. Vorsorge statt Nachhaltigkeit – Ethische Grundlagen der Zukunftsverantwortung. Pp. 19–56 in *Zukunftsverantwortung und Generationen solidarität*, edited by D. Birnbacher and G. G. Brudermüller. München: Königshausen und Neumann.

Blackbourn, D. 2008. *Die Eroberung der Natur: Eine Geschichte der Deutschen Landschaft*. München: Pantheon.

Bloch, E. 1980. *Geist der Utopie*. Frankfurt: Suhrkamp.

Bogdanow, A. 1989. *Der rote Planet*. Berlin: Volk und Welt.

Brandt, D. 2007. *Der deutsche Zukunftsroman 1918–1945: Gattungstypologie und sozialgeschichtliche Verortung*. Tübingen: Max Niemeyer Verlag.

Brimblecombe, P. 2004. Apokalypse im Nebel. Der Londoner Smog im späten 19. Jahrhundert. Pp. 42–51 in *Wird Kassandra heiser? Die Geschichte falscher Ökoalarme*, edited by F. Uekötter and J. J. Hohensee. Wiesbaden, Stuttgart: Franz Steiner Verlag.

Callenbach, E. 1996. *Ecotopia: The notebooks and reports of William Weston*. Stuttgart: Reclam.

Dürnberger, C. 2008. Der Mythos der Ursprünglichkeit–Landwirtschaftliche Idylle und Ihre Rolle in der öffentlichen Wahrnehmung. *Forum TTN* 19: 45–52.

Dürnberger, C. 2012. Conflict cloud green genetic engineering: Structuring and visualizing the controversy over biotechnology in agriculture. Pp. 427–30 in *Climate change and sustainable development. Ethical perspectives on land use and food production*, edited by T. Potthast and S. Meisch. Wageningen: Wageningen Academic Publishers.

Fest, J. C. 1991. *Der zerstörte Traum: Vom Ende des utopischen Zeitalters*. Berlin: Siedler.

Fleck, D. C. 2009. *Das Tahiti-Projekt*. München: Piper.

Foucault, M. 1993. Andere Räume. Pp. 34–46 in *Aisthesis: Wahrnehmung heute oder Perspektiven einer anderen Ästhetik – Essais*, edited by K. Barck. Leipzig: Reclam.

Freystedt, V., and Bihl, E. 2009. *Equilibrismus: Neue Konzepte statt Reformen für eine Welt im Gleichgewicht*. Wien: Signum Verlag.

Gill, B. 2003. *Streitfall Natur: Weltbilder in Technik- und Umweltkonflikten*. Wiesbaden: Westdeutscher Verlag.

Habermas, J. 1985. *Die neue Unübersichtlichkeit. Kleine Politische Schriften V. Neue Folge*. Frankfurt: Suhrkamp.

Hölscher, L. 1985. Der Begriff der Utopie als historische Kategorie.Pp. 402–18 in *Utopieforschung: Interdisziplinäre Studien zur neuzeitlichen Utopie*, edited by W. Voßkamp. Frankfurt: Suhrkamp.

Jonas, H. 1984. *Das Prinzip Verantwortung. Versuch einer Ethik für die technologische Zivilisation*. Frankfurt: Suhrkamp.

Korthals, M. 2003. Grüne Gentechnik. Pp. 354–62 in *Bioethik: Eine Einführung*, edited by M. Düwell and Steigleder, K. Frankfurt: Suhrkamp.

Marcuse, H. 1980. *Das Ende der Utopie*. Frankfurt: Suhrkamp.

More, T. 2016. *Utopia*. London/New York: Verso.

Müller, C. 2011. *Urban Gardening: Über die Rückkehr der Gärten in die Stadt*. München: oekom.

Popper, K. 1992. *Die offene Gesellschaft und ihre Feinde. Band 1. Der Zauber Platons*. Tübingen: UTB.

Radisch, I. 2010. Das Paradies auf Erden – Was ist von der Lebensreform geblieben? Zu Besuch in der einstigen Gartenkolonie Eden. Pp. 72–80 in *Zeit Geschichte Magazin. Das Deutsche Kaiserreich*, Vol. 4. Hamburg: Zeit-Verlag.

Radkau, J. 2002. *Natur und Macht: Eine Weltgeschichte der Umwelt*. München: Beck.

Ricoeur, P. and Greisch, J. 1996. *Das Selbst als ein Anderer*. München: Wilhelm Fink.

Rifkin, J. 2000. *Das biotechnische Zeitalter: Die Geschäfte mit der Gentechnik*. München: Campus-Verlag.

Rüsen, J. 1985. Utopie und Geschichte. Pp. 356–74 in *Utopieforschung. Interdisziplinäre Studien zur neuzeitlichen Utopie*, edited by W. Voßkamp. Frankfurt: Suhrkamp.

Saage, R. 1995. *Vermessungen des Nirgendwo: Begriffe, Wirkungsgeschichte und Lernprozesse der neuzeitlichen Utopien*. Darmstadt: Wissenschaftliche Buchgesellschaft.

Saage, R. 2002. *Utopische Profile: Industrielle Revolution und Technischer Staat im 19. Jahrhundert*. Berlin: Munster.

Schäfer, L. 1996. Die Idee der zu schonenden Natur. Pp. 199–226 in *Naturauffassungen in Philosophie, Wissenschaft, Technik: Gegenwart*, edited by L. Schäfer and E. Ströker. Freiburg: München: Alber.

Schwonke, M. 1957. *Vom Staatsroman zur Science Fiction, Eine Untersuchung über Geschichte und Funktion der naturwissenschaftlich-technischen Utopie*. Stuttgart: Ferdinand Enke Verlag.

Seibt, F. 1985. Utopie als Funktion abendländischen Denkens. Pp. 254–79 in *Utopieforschung. Interdisziplinäre Studien zur neuzeitlichen Utopie. Erster Band*, edited by Wilhelm Voßkamp. Frankfurt: Suhrkamp.

Sieferle, R. P. 1984. *Fortschrittsfeinde? Opposition gegen Technik und Industrie von der Romantik bis zur Gegenwart*. München: Beck.

Thoreau, H. D. 2009. *Walden oder Leben in den Wäldern*. Köln: Anaconda.

Wöbse, A. K. 2005. Zur visuellen Geschichte der Naturschutz-und Umweltbewegung: Eine Skizze. Pp. 222–46 in *Natur- und Umweltschutz nach 1945: Konzepte, Konflikte, Kompetenzen*, edited by F. J. Brüggemeier and J. I. Engels. Frankfurt: Campus.

15 'Growing your own'

Gardens, plants, and the Good Life

David E. Cooper

For many people, gardens, plants and food form an inseparable trio. They garden, in part at least, in order to grow vegetables or harvest vegetables, fruit and nuts for the purpose of eating them. This chapter reflects on this trio and its relevance to the good life, and it does so in the context of the emergence of 'plant ethics' in recent years.

Plant ethics

In fact, it is misleading to refer to the recent emergence of plant ethics because while the label is new, there is nothing new about moral thinking that attends to plants. The Confucian philosopher, Mencius, famously described human nature as containing shoots that, properly tended, sprout into compassion, wisdom and other forms of goodness. For Chinese thinkers more generally, one scholar observes, plants and water are the 'root metaphors' for their conception of human life (Allan 1977, 100, 133f). In the West, Rousseau's (1974, 6) comparison of a child's education with the cultivation of a plant was to inspire two centuries' of 'child-centered' pedagogy. Nor is it new to express concern for how plants are treated. Japanese Buddhist texts, for example, encourage the idea that vegetal, as well as animal, life are if not sentient, then, at any rate, possessed of 'Buddha-nature' and hence deserving of compassion and protection from 'violence' (James 2004, 64–72). Finally, there is nothing new in the idea that proper attention to plants is good for human beings: for Greek and Roman writers, including Xenophon and Virgil, cultivating the soil and tending the garden was not simply healthy and enjoyable but also 'pious' and pleasing to the gods (Ray 2010). In addition, for many writers of ancient Greece and Asia, mindful attention to plants is morally educative: plants set examples – of sturdiness, patience, dignity – that we do well to emulate.

What is novel is the *idiom* of much of the literature of contemporary plant ethics. Unsurprisingly, this idiom reflects the prevailing tendencies in environmental ethics and modern moral philosophy more generally. In particular, one finds in this literature an emphasis on 'rights', 'intrinsic value', 'justice', 'respect' and 'dignity'. A good example of the literature is the 2008 report of the Swiss Federal Ethics Committee on Nonhuman Technology. Transparently titled

'The Dignity of Living Beings with Regard to Plants', it exhorts us to show 'moral consideration of plants for their own sake' and to recognize the moral dignity of plants, their 'absolute moral value', and rights (cited in Marder 2013a, 179f). Books on permaculture also typically employ this idiom. The author of one of these, for example, refers to the 'intrinsic right to life' that pine trees and blades of grass possess (Bell 2004, 29).

It would be unfortunate and strait-jacketing to restrict plant ethics to enquiries that conform to the idiom of rights, intrinsic value and related notions. Indeed, below, I will suggest that this vocabulary is overly abstract, idle and redundant. Even if that is too harsh a judgment, it is surely the case that we should be addressing a very general question that is not obviously, at least, best addressed in the contemporary idiom. This is the question of how human beings should relate to plants. What is a good relationship between the two? How should a person conduct his or her life with respect to plants? How might plants contribute to the good life?

These are indeed very general questions, and there is merit in narrowing them down to questions that are more tractable. Since one of the most important ways in which plants figure in our lives is by being eaten, a sensible focus might be on plants as food. However, that is still a very broad focus, one that generates many different kinds of questions. These include questions about the permissibility of genetically modified crops, the objections to mono-cultural agriculture, and the impact of modern agriculture on indigenous peoples. Thus, let us narrow the question further by taking it in conjunction with another important way in which plants enter into people's lives – namely, as occupants of their gardens (including allotments). Nothing, perhaps, very much turns on focusing our question on food gardening: most of what I will say with suitable amendments, applies to other areas of gardening, such as growing roses. However, having a particular area in mind helps give purchase, in later sections of the chapter, to the discussion of virtues and understanding. The focus on 'growing your own' also has the advantage of discouraging the unfortunate impression – encouraged by an overblown analogy between gardening and painting – that the garden is essentially, or first and foremost, an aesthetic spectacle.

Our focus, then, will be on the area where plants as food and plants as occupants of gardens overlap or intersect: in effect, on the trio of gardens, plants and food. Thus, our question becomes one such as the following: 'How does "growing your own", in your garden, constitute a good relationship to plants, to vegetal life?', or more simply, 'Why is it a good thing for you to grow your own?'. The intention is that the answers essayed to these questions will illuminate wider issues about our relationship to plant life, and possibly to nature at large.

'Global' and 'Personal'

The questions just formulated may be narrower than the previous questions, but they have invited many answers, for example in works on permaculture. The author of one of these works divides the answers into two broad but convenient groups: those that appeal to 'planetary' and 'personal' reasons for growing your own (Burnett 2014, xi). 'Planetary' reasons for growing your own are such umbrella objectives as 'greening the planet' and 'saving the earth'. If all or most of us grew our own, then, it is argued, there would be a massive reduction in transport costs, deforestation, the demand for water, the use of chemicals, and much else that is deemed responsible for global warming and other environmental ills. Our 'ecological footprints' would shrink and the Earth be given a chance to recover from the technological onslaught that the modern food industry has unleashed.

I am not very interested in these planetary reasons since I do not see that they are reasons why you or I should grow our own, or more generally why growing one's own enhances a person's life and is a constituent of the good life. That it would be to the benefit of the planet and the human race if most people grew their own has no immediate relevance to what you or I should do, since we know very well that most people will not grow their own. To take an analogy, suppose that Utopian communists, such as Levellers and Cabetistes, were right to think that the world would be a better place if everyone renounced private ownership. That is no reason for me to give away my house and savings in the world as it actually is, one in which hardly anyone voluntarily gives up what he or she owns. Arguments taking the form 'If everyone's doing something would be good, then you should do it', despite their popularity among some schools of utilitarianism, are without force.

It is worth adding, perhaps, that anyway, planetary arguments rely on total speculation. We surely know that a world in which nearly everyone grew their own would be a very different world from the actual world – the result, perhaps, of some catastrophe: a nuclear war, a collision with a meteorite, a new ice age, or whatever. (It was during the relative catastrophe of the Second World War that people in large numbers 'dug for Britain' and grew their own.) The question of what would be beneficial or wise to do in such an unknown world is an idle one.

Therefore, my interest lies in 'personal' reasons, ones for you or me to grow our own and that do not invoke speculative scenarios.[1] Many of these personal reasons straightforwardly indicate the economic, physical and gustatory benefits of growing your own. Digging a vegetable patch keeps you fit, saves you spending money at the supermarket, and provides you with fresh, tasty produce. These, however, are not the kinds of benefits, perhaps, that enquiry into how gardening may contribute to the good life seeks to identify. They are, arguably, too straightforward and obvious and without the moral charge with which the idea of the good life is invested. However, one needs to tread carefully here. In my judgment, it is wrong to look for a sharp divide between moral benefits and, say, bodily or economic benefits. Roman writers, and their neo-classical admirers, such as Thomas Jefferson, emphasized the sturdiness, self-sufficiency and independence of the hard-working farmer or kitchen gardener (Adams 1991, 287). Are these moral goods or goods of a different order? The same question can be asked of the goods that Chinese thinkers associated with bodily disciplines, including gardening: 'keep your body whole' so that you may eschew an obsession with wealth and fame, and come into harmony with 'the single energy that is the world' (Ziporyn 2009, chaps. 5 and 22).

Still, one sympathizes with the desire to identify goods more traditionally associated with 'the good of the soul' than muscle tone and the tastiness of fruit. What we are looking for, I suggest, is an account of how growing your own may also be a cultivation of certain virtues and of an implicit understanding of the way of things. The virtues and understanding of reality were, of course, the two inseparable goods that marked the lives of the sages, enlightened persons and saints of ancient teachings – the two great goods whose combination was called 'wisdom'. In the final analysis, it is not going to matter whether we attach the label 'moral' to the virtues and understanding that gardening may cultivate. What matters is to recognize that these are ingredients in a human life that flourishes, is lived well, and is in a right relationship to the living world.

Virtues

That the garden is a place in which to practice virtues was itself a familiar idea in various ancient traditions. To cultivate his garden, remarked Pliny, was to cultivate himself (1963, 112). It is an idea that has been repeated over the centuries, for example by Victorian enthusiasts for whom a hard day's digging was an effective antidote to illicit sexual desires. However, as that example suggests, the idea is a very diffuse one, and it is important to distinguish between merely contingent and essential connections between gardening and the virtues. Gardening as a cure for preoccupation with sex is an example of the former, as is the role of the garden as a hospitable place for edifying conversations with friends or for what Andrew Marvell found equally edifying – 'delicious solitude'. So is the fact, if it is one, that because a morning's pruning makes me feel both relaxed and fresh, I am likely to be nicer to people than I usually am.

To grasp what an 'essential' connection between gardening and virtue might be, it is helpful to emphasize that gardening is a *practice* or, as the Japanese call it, a *way* (*dō*). Practices, in the serious sense urged by MacIntyre (1982) are complex, goal-directed activities to which certain virtues are 'internal'. That is, practices cannot be properly and successfully engaged in unless the practitioners exercise these virtues. The connection between growing your own and certain virtues, therefore, is not merely contingent: it is not accidental in the way that my pruning and my being nice to people are.

It would have to be an exceptionally lucky kitchen gardener whose fruit trees and vegetable patch flourished despite his extreme impatience, his lack of care or concern for the good of the plants, his rigidity and lack of responsiveness to changing conditions, his refusal to recognize that his success is dependent on many factors in addition to his own work, and his complete insensitivity to the fecund beauty of the fruits and plants that grow.

This reflection suggests a way of identifying the virtues internal to the practice of growing your own. They will be the opposites of the failings of the gardener I just described. They will include patience, care and concern for living things, mindfulness and flexibility, humility, and appreciation of beauty. These virtues are attitudes and dispositions without which the gardener would have to be *very* lucky to succeed. This does not, perhaps, amount to saying that these virtues are 'essential' or 'necessary' in the sense that metaphysicians usually have in mind. It is not a matter of logical or cosmic impossibility. One can, after all, just about *imagine* a gardener succeeding despite lacking these virtues: maybe, where he lives, nature is so bountiful and benign that his garden succeeds despite his failings, or perhaps some local *kami* or nature spirits come to the rescue. However, we would hardly want to say that the connection between virtues such as care or humility and horticultural success is accidental or 'merely' contingent.

Part of the answer, then, to the question of how growing your own may be a constituent of the good life and represent a right relationship to the living world is the following: properly or authentically engaged in, growing your own is a practice, a 'way', to which a number of virtues are internal or essential. In learning how to grow your own, you must also learn how to exercise these virtues in relation to living things: for this reason, to recall Pliny's words, cultivation of the garden is also a cultivation of the self.

Some people may react to this by insisting that at least some of the virtues of gardening that I have mentioned are not really virtues – not, at any rate, if virtues are *moral* qualities. Appreciation of beauty, for example, is not a virtue, according to this criticism. The reply to the criticism was anticipated when I wrote that, in the final analysis, it does not matter much whether the term 'moral' is applied or not. It just depends on how the term is taken. Patience, say, or aesthetic sensibility is unlikely to count as a moral requirement on a distinctively modern understanding of morality as a system of principles governing actions. However, we should remind ourselves that moral or ethical reflection was once understood less narrowly, as reflection, in effect, on the good life. In my judgment, it would be difficult to deny that appreciation of the beauty of trees, bushes, plants and fruit crucially belongs to a relationship to the living world that it is good to promote and, hence, to the good life itself.

Understanding

That the garden is a place in which to develop understanding – of nature and its relationship to human culture, included – is also an ancient conviction. However, the word 'also' is perhaps misleading. The reason, one commentator explains, why Epicurus's garden was a site for moral 'self-cultivation' was that it also provided 'a form of education in the ways of nature, in the 'greater harmonies' of the cosmos and the soul's 'connection to matter' (Harrison 2008, 73–74). More generally, cultivating the virtues in the garden was also to be acquiring understanding or, as the Daoists put it, to be embarking 'on the Way'. This should come as no surprise, given the unity of virtue and knowledge – of their union in wisdom or enlightenment – that was central to many ancient dispensations, Eastern and Western alike.

In light of this doctrine of unity, it is useful to consider the manner in which various of the virtues of growing your own are intimately connected with forms of understanding. One of these virtues was humility. The virtue of humility is not a reticent feeling of modesty, let alone a kind of self-contempt. Rather, it is the recognition that even the most creative and energetic of gardeners are dependent for the success of their endeavors on what they themselves can take no credit for. This will be, above all, the cooperation of nature. The healthy squash that Pollan (1996, 73) grew and wrote about was as much a 'gift' that owed to nature's 'grace' as the product of his own labor. Another author more widely refers to plants as 'constantly giving gifts to sustain human life' (Hall 2010, 46). Humility is inseparable, then, from an understanding of the dependence of human practices, indeed human culture, on nature. However, humble gardeners will also acknowledge their dependence on traditions of growing, on a folk garden wisdom that they draw upon, and even on what they have absorbed from horticultural TV programs and magazines. The humble gardener, one might say, knows his or her place in the human, as well as the natural, world.

Consider next the virtue of what I called mindful flexibility and responsiveness. By this, I do not mean, simply or primarily at least, a sort of nimble alertness of the kind that a sensible person or animal will display when, for example, scenting danger. Rather, I intend what the Daoists refer to as 'spontaneity' (*ziran*), understood as a lucid openness to how things, including plants, themselves are and a readiness to adjust how one treats them in response to this openness. Crucially, this requires sensitivity to the good of the plants, of what it is for them to flourish. This is a requirement that militates against artificial perceptions of plants dictated by conventions, fashions and prejudices – including those that often govern the awarding of prizes at horticultural shows. It also militates against 'passions which distort awareness' (Graham 2001, 14) and perceptions and understanding of plants that are entirely subjected to the utility that living things have for us. That the apple tree was planted in order that a family might eat its fruit does not warrant doing whatever the family likes to maximize its yield, irrespective of the needs of the tree.

Finally, consider appreciation of the beauty of apple trees, climbing beans, ripening aubergines and the like. This should not be taken in a thin way as simply some pleasurable sensory experiences caused by certain visual or olfactory stimuli. It is not an enjoyment of 'appearances' but an enjoyment that registers and requires recognition and understanding. Marcus Aurelius was right to argue that natural beauty is not confined to things that 'a man of taste' would necessarily enjoy and that it is properly available only to people who have 'familiarized [themselves] with nature and its workings'. Appreciation of the beauty of what might at first seem an unattractive plant or animal is enabled once we understand the place that it has in the wider processes of nature or the functions that its various parts – knobbly excrescences, waving tendrils, 'gaping jaws', or whatever – perform in the overall economy of the plant's or animal's life (Aurelius 1983, Bk 3 §2). It is when we recognize and, as it were, 'feel' the fruit tree's rootedness in the soil, its hospitality to the birds and wasps that visit it, its exploitation of the sun and the rain, and its adaptability to the seasons that a deep appreciation of its beauty is born. In addition, this is an appreciation further enhanced by also recognizing the marvelous congruence of form and function that is manifest in the fruit tree. It inflects our appreciation of the delicate springtime blossoms on the fruit tree to see them not just as pretty colored shapes but also as presaging the ripe fruit that, in autumn, will grow where they once bloomed. Perhaps this is one reason why Japanese men and women take their quiet pleasure in the blossoms that fall from plum and cherry trees.

I could continue pointing out the involvement of understanding – of nature and its relation to culture – in the exercise of the virtues presupposed by the practice of growing your own. However, the virtues that I have discussed are sufficient to establish that there is this involvement. There is much more that could be said, more generally, about the linked roles of virtue and knowledge – of wisdom – in gardening, something attested to by a burgeoning literature (see, e.g. Brook 2010; Cooper 2012, 2015; Di Paola 2015, 2017). However, the contours, at least, of an answer to the question of how growing your own may be a constituent of the good life, I hope, are clear. Good gardening is a practice that calls upon, cultivates and exercises various virtues. In doing so, it invokes and manifests a live, practical understanding of important truths about the living world and the relation of that world to human culture. To that extent, good gardeners are both virtuous and 'in the truth'. These are excellent 'personal' reasons for you and me to grow our own. Our doing so will not 'save the planet', but our lives will have been enhanced.

Plant ethics again

In the first section of this chapter, I briefly mentioned the contemporary interest in plant ethics, and indicated that I would return to considering the discussions that this interest has generated. I do so less out of scholarly interest than in order to gauge the bearing of these discussions on the approach I have been defending in the chapter. Are they tangential to my approach, or may they, instead, modulate and deepen the claims about gardening, virtue and understanding that I have defended?

The unsurprising answer is that it depends on the particular direction that the discussions take. However, before proceeding to that, it is worth noting some themes and emphases that are common to these discussions. First, there is the insistence that plants are 'living beings'. They do not simply grow and, until they die, remain alive: like animals and people, they *have* lives. The point is often put in terms of the impressive degree of autonomy, self-direction and strategies for survival that plants manifest. Second, relatedly, it is held to be not only legitimate but also compelling to employ a psychological vocabulary of intentionality to describe plants, including terms such as 'purpose', 'intelligence' and even 'mind'. Typically, this claim is immediately qualified by rejecting any attribution of 'consciousness' to plants. The 'thinking' of plants, as Marder (2013b, 124) puts it, is 'non-conscious', a 'thinking without the head'. Mabey (2015, 337) also defends a notion of intelligence uncoupled from self-awareness Third, in the light of this recognition of the autonomy and intelligence of plants, it is argued, plants deserve to be accorded a 'respect' that instead, in the Western tradition at least, has been markedly absent.

However, it is at this point – with the turn towards the ethical implications of the proper understanding of plant life – that differences emerge. For some authors, such as the permaculturalists and the authors of the Swiss Committee Report referred to above, respect for plants is essentially recognition of their moral status. They are beings with moral 'rights', 'dignity' and 'intrinsic value': for this reason, they should be given respect and treated accordingly. My own account of a good relationship to plants was, by contrast, free from appeals to such concepts, and I am out of sympathy with the 'two-stage' idea that in order to articulate a good relationship to living beings, we must first determine their 'moral status' (see Zahir 2007, chap. 2).

'Out of sympathy' is, perhaps, too mild because I find the rhetoric of plant 'rights', 'intrinsic value', and the rest excessively abstract and idle, as well as redundant. It is excessively abstract since, by talking of the value of plants in general, it ignores the particular contexts and the concrete engagements that give force to speaking of a plant's value. It is, for example, in the context of a garden that a certain tree may be one to honor and value. The value that an apple tree planted in celebration of a marriage or a birth has for a family or that a grape vine that has survived centuries of storms has for a community is not one that is shared by a tree in a remote forest that no one has ever seen. That the rhetoric is also idle owes, in part, to its abstract character. It is idle because it is entirely unclear what, if anything, turns on the rights and value allegedly possessed by all plants. What, if anything, is implied for how plants should be treated? Authors who attribute 'an intrinsic right to exist' to plants do not, it seems, want to exclude the eating of plants (Bell 2004, 24). So, what is excluded? 'Arbitrary harm' to plants, we are told, must be avoided, but this simply raises the question of what counts as 'arbitrary'. It becomes difficult, in effect, to see what the rhetoric of rights and values adds to the bare and vague call to respect plants. If it adds nothing, then the invocation of the 'moral status' of plants has not played any role at all.

This is one reason for regarding the rhetoric as redundant, but it is important to also recognize that in order to discern how we should relate to plants, there is anyway no need for the 'two-stage' detour via moral status and rights. This is so, at any rate, if you are sympathetic to my account of the exercise of virtues and understanding as constituting a good relationship to plant life. The good gardener is someone for whom, seeing plants for what they are – in their place in their environments, in their beauty – there is no question as to how he or she should attend to and engage with them. It belongs to the virtue of spontaneity – one that is itself promoted in the garden – to respond appropriately to plants. No gardener is likely, before responding, to pause and ask 'But do they have intrinsic value?'.

However, there are writers who take a different direction since they either eschew the vocabulary of moral status altogether or express suspicion of it. The metaphor in the title of Mabey's (2015, 338) *The Cabaret of Plants* – a book that avoids the language of rights and moral value – is a rich one. By emphasizing the 'show' that plants put on for us, the metaphor militates against the reduction of plants to 'aggregates of neurotransmitters and signalling network'. Yet the plant show is not a mere aesthetic spectacle of which we are passive observers. A cabaret is a show with a 'participating audience', and that is what we are with respect to plants. We should be 'natural co-operators' with them and seek to nurture an 'intimate, interactive relationship between our two spheres of existence', human and vegetal (Mabey 2015, 8).

The same emphasis on engaged relationship is found in Marder's (2013a, 180) writings. Although he flirts with talk of the 'moral dignity' of plants, he is generally suspicious of such terms: to talk of plants as 'moral *objects*', for example, is to encourage a view of them as standing entirely outside the human sphere and, hence, ironically, as being at our disposal. We display a 'genuinely ethical' attitude towards plants not by accepting an 'obligation' towards them but by a 'cultivation of intimacy' with them, of a 'full-fledged relation' of the sort that Martin Buber had in mind by the 'I-Thou' relation (Marder 2013a, 187). An 'ethical approach' to plants is manifested not in 'a quasi-religious veneration' of them, and even less in an abstract acknowledgment of their moral status, but in practices that represent a 'refusal to regulate the human relation to plants on the basis of commodity-economic logic' (Marder 2013a, 185), for example in the eating of locally grown fruit and vegetables – or, one might surely add, in eating home-grown ones. This is not because local or home-grown produce is less expensive or tastier but because to care about the food's provenance and one's connection to it is already a refusal to submit entirely to a 'commodity-economic logic' (see also Di Paola 2017; Purdy 2015).

For Mabey and Marder, then, plant ethics is not a theory of the rights or moral value of vegetal life from which principles of conduct are then derived but a reflection on enlightened forms of intercourse and engagement with it. It is the kind of reflection that is the mark of the good gardener but signally absent, one imagines, from the mind set of an industrial scale tomato growing enterprise.

Conclusion

Unlike the approach to plant ethics that focuses on the moral status of plants, one that instead focuses on our engaged relationship with them is entirely consonant with ancient moral wisdom. Xenophon, Virgil, Mencius, and Zhuangzi, for example, saw no need – indeed, would have lacked the conceptual resources – to confer moral rights or intrinsic value on plants, or indeed animals and people, in order to appreciate that there are good and bad ways of relating to them. The cultivation and exercise of virtues and understanding that growing your own requires and promotes is precisely the kind of relationship that I take to be endorsed by Mabey's and Marder's allusions to an 'intimate, interactive' or a 'full-fledged' and 'co-operative' – even an 'I-Thou' – relationship to plant life.

The best writing on plant ethics may even help to inform and encourage virtues such as humility, care and the appreciation of beauty and the forms of understanding that these virtues invoke. Mabey's book, in particular, is full of detailed, concrete and charming descriptions of vegetal life that, without any assistance from 'moral theory', make it quite clear how incompatible with the good life it would be to treat plants as nothing but a 'standing reserve', as Heidegger (1977, 17) put it, that is at our disposal. Much of what Mabey describes will already be known, implicitly and, as it were, 'in the hands', by good gardeners. If I am right, this knowledge is essential to being a good gardener, someone whose work is integral to a good life.

Note

1. One work in the philosophy of gardens that plausibly conjugates "personal" and "planetary" arguments is Di Paola (2017).

References

Adams, W. H. 1991. *Nature perfected: Gardens through history*. New York: Abbeville Press.

Allan, S. 1977. *The way of water and sprouts of virtue*. Albany, GA: State University of New York Press.

Aurelius, M. 1983. *The meditations*. Indianapolis, IN: Hackett.

Bell, G. 2004. *The permaculture way: Practical steps to create a self-sustaining world*. Hampshire, UK: Permanent Publications.

Brook, I. 2010. The virtues of gardening. Pp. 13–25 in *Gardening: Philosophy for everyone*, edited by D. O'Brien. Oxford, UK: Wiley-Blackwell.

Burnett, G. 2014. *The vegan book of permaculture*. Hampshire, UK: Permanent Publications.

Cooper, D. E. 2012. Gardens and the way of things. Pp. 20–33 in *Earth perfect: Nature, Utopia and the garden*, edited by A. Giesecke and N. Jacobs. London: Black Dog.

Cooper, D. E. 2015. Gardeners of the cosmos: The way of the garden in East Asian tradition. Pp. 38–49 in *The good gardener? Nature, humanity and the garden*, edited by A. Giesecke and N. Jacobs. London: Artifice.

Di Paola, M. 2015. Virtues for the anthropocene. *Environmental Values* 24, no. 2: 183–207.

Di Paola, M. 2017. Ethics and politics of the built environment. Gardens of the anthropocene. Dordrecht, NL: Springer.

Graham, A. C. 2001. *Chuang-Tzu: The Inner Chapters*. Indianapolis, IN: Hackett.

Hall, M. 2010. Escaping Eden: Plant ethics in a gardener's world. Pp. 38–47 in *Gardening: Philosophy for everyone*, edited by D. O'Brien. Oxford, UK: Wiley-Blackwell.

Harrison, R. P. 2008. *Gardens: An essay on the human condition*. Chicago, IL: University of Chicago Press.

Heidegger, M. 1977. *The question concerning technology and other essays*. New York: Harper & Row.

James, S. P. 2004. *Zen buddhism and environmental ethics*. Aldershot, UK: Ashgate.

Mabey, R. 2015. *The cabaret of plants: Botany and the imagination*. London: Profile.

MacIntyre, A. 1982. *After virtue: A study in moral theory*. London: Duckworth.

Marder, M. 2013a. *Plant-thinking: A philosophy of vegetal life*. New York: Columbia University Press.

Marder, M. 2013b. What is plant-thinking? *Klesis: Revue Philosophique* 25: 124–43.
Pollan, M. 1996. *Second nature: A gardener's education*. London: Bloomsbury.
Purdy, J. 2015. *After nature: A politics for the anthropocene*. Cambridge, MA: Harvard University Press.
Ray, M. T. 2010. Cultivating the soul. Pp. 26–37 in *Gardening: Philosophy for everyone*, edited by D. O'Brien. Oxford, UK: Wiley-Blackwell.
Rousseau, J.-J. 1974. *Émile*. London: Everyman.
Zahir, T. 2007. *Ethics and the beast: A speciesist argument for animal liberation*. Princeton, NJ: Princeton University Press.
Ziporyn, B. 2009. *Zhuangzi: The essential writings with selections from traditional commentaries*. Indianapolis, IN: Hackett Publishing Company.

16 "Hey Plants, Take a Walk on the Wild Side!"

The ethics of seeds and seed banks

Nicole Karafyllis

Introduction

Lou Reed's famous 1972 song "Take a walk on the wild side" is about homosexual identities intertwined with urban ethnographies, and about the coming out of "wild" or authentic selves. These selves, however, can only exist off-site, or *ex situ*, that is in restricted times and spaces far from the times and places of origin. The "wild side" is in a parallel world, as it were. Interestingly, in both the world of origin and the world of the "wild side", the self is only a *temporary* self. In the end, the artificial practices of "being wild" that the imaginary protagonists of Reed's song engage in kill or impair the self and its co-selves.

Comparable ambivalences are the topic of this chapter. Let us imagine that Reed's song is about plants, and more specifically crops kept in the parallel world of high-tech seed banks. This chapter intends to reflect on seed banks and their links with ethical challenges. It shows how the potential of seeds for (metaphorically speaking) "walking on the wild side" are ontologically, epistemologically and technically modelled. I will focus on the modes of potentiality and continuity of plant populations, which are embodied by seeds. For comparative reasons, the analysis will reach back to interpretations of "seed" given by Aristotle and Theophrastus, contrasting them with Heidegger's concept of "stock" ("Bestand"). I will close by summarizing the reasons why a closer look at seed banks is of general relevance to plant ethics.

It is useful to begin by introducing seed bank-related terms and concepts. Seed banks, like botanical gardens and zoos, belong to the category of "living collections" (Karafyllis 2018). Because the objects collected are alive, living collections differ from natural history collections (Kohler 2006). Living collections work on the assumption that "life" is collectable and, as such, can be *kept alive* under storage conditions.

Since the 1970s, seed banks have changed their names to "gene banks" and "plant germplasm systems" (Karafyllis and Lammers 2017). During the last century, the object of collection has become ontologically vague: by now it can range from seed to germplasm, from gene to genome.[1] However, the focus on resources and technological systems functions as a unifier. This is partly due to the FAO concept of "plant genetic resources", which has gained popularity since the 1960s

(Pistorius 1997). Theoretically, the concept applies to all plants. Following the OECD criteria of the new millennium (OECD 2001), some of the bigger seed banks have recently converted into "plant Biological Resource Centers" (pBRCs). This label indicates boundaries between plant and microbial domain (mBRC), and animal domain (aBRC). These labels all suggest that the entirety of a biological domain can be conserved *ex situ*. This may be wishful thinking.

In seed banks, seeds are allowed to become actual plants only approximately every 30 years, when the curator is rejuvenating the seeds in order to sustain their viability. Once in a blue moon, then, banked seeds can "take a walk on the wild side". After harvest, the next seed generation is jarred again and transferred to the cooling chamber, where it will be put to sleep for the following decades. Of course, useful plants, and among them predominantly crops, get most of the attention of seed banks and constitute the majority of accessions[2] (i.e. aggregated seed samples) worldwide.

Much like Reed's "wild side" is situated in the very mannered arty context of 1970s New York, "the wild" of banked seeds that are allowed to burst into plants is a controlled environment: the acre, the garden, or the greenhouse. These environments are less controlled than a modern seed bank is, and much like in Reed's song it is not easy to tell what environment is riskier for the seeds of a plant population – the "wild" or the "controlled".

When a seed of a wild plant arrives at the soil surface, it commonly persists in the soil for a certain time span. In a "soil seed bank", conditions "may vary temporally from year to year and spatially within a year, and may reflect both abiotic variables (chiefly climate) and the presence and activities of other plants and animals." (Thompson 2000, 218) For seeds, persisting in the soil is both useful and risky. On the one hand seeds that fail to germinate at the first opportunity may die before they are able to do so, but on the other hand the early seedling may die because of unfavourable conditions. Consequently, the reproductive fitness of the population would be reduced. A soil seed bank, with its different times of persistence for individual seeds functions as a buffer system to maximize the population's survival. Further adaptive significance have the different types of seed dormancy, which vary among species.

By contrast, the seeds of cultivars have been technically adapted to agriculture. This affects seed dormancy, persistence, and more. Since agriculture and breeding began about 10,000 years ago, peasants have been synchronizing the ripening times of fruits, including seeds. In the case of wild plants, time-spans naturally differ already on each individual plant, in order to increase the population's reproductive fitness (see above). Furthermore, a continuous breeding goal is to keep the fruits/seeds at the stage of infructescence. In the case of cultivated cereals, for example, the ripe grains do not fall off onto the soil like the fruits of grasses do (grasses are the ancestors of cereals). The ripe grains of cereals "unnaturally" stick to the ear, which prevents fruit/seed loss before harvest. Strictly speaking, the seed of a cultivar is no longer a *dispersal unit*. In order to become a next generation of plants, the seeds need to be harvested (and then sown). This agricultural, rather than botanical, practice treats seeds as final products of a production process. This leads us to think of seed banking as related to harvesting. At the same time, it complicates the application of ethical approaches that emphasize plants' *continuity* and *potentiality*, which necessarily refer to beginnings and origins (see below).

Plant ethicists might want to say that by means of generating seeds, plants show their interest in their own continuation (as individual, population or species). Agriculture runs against such interest, as it can be described as a technological process of *dis-continuation* of plant populations successions and *selection* of favourable traits, picked up from the plant's substantial potentiality (which we might call "wildness"). Seed banks' practices can be similarly described. Yet environmentalist discourses on rapid climate change, loss of biodiversity, and land degradation tend to portray plants as being at risk not just in the wild but also in their cultivated worlds, that is as crops grown on acres. To that extent, they seem to encourage the modern practice of banking seeds. Seed banking then counts other factors as risks, including the sexuality of flowering plants, as we will later observe.

To understand how a cultivar can still "walk on the wild side" and why this ethically matters, we first need to understand the key concept of "wild". "Wild plant" can denote either a plant "in wilderness" – grown apart from human interventions in areas where few humans are to be found – or a plant's "wildness" – the autonomy of its growth and growth potentials. "Wild" in the latter sense may, for instance, refer to the level of breeding status, which can range from that of a low-bred landrace to a high-bred crop obtained by means of genetic engineering. On a commonly held view, influenced by environmental ethics, "wild" plant populations are inherently more valuable and informative than their cultivated counterparts. "Wild", here, typically means "in wilderness" (somewhere "outside" the world of humans). This fosters non-anthropocentric thinking. Yet the majority of plant conservation strategies worldwide address plants for food and agriculture. Consequently, if plant ethics wants to make a real difference, it will have to argue from the perspective of anthropocentrism, and take into full account human culture and technology. For one thing, what the "wild side" can be for cultivars is not immediately clear. This should also remind us that cultivated plants should not be thought of in analogy with domesticated animals, as though a plant could be "set free" (cf. Rolston 2004). Generally, the legal frameworks presiding over seed banking include ethical concepts that refer both to wilderness (and thus tie to the idea of *in situ* origins) and to the value of plants' wildness, that is their autonomous becoming, which may well take place on acres or in seed banks, at least partially.

When referring to "a plant" or "a seed", I usually address the plant population or its collected samples in the seed bank, that is the *model* of a plant population. Generally speaking, the plant population (and its banked model) indicates varieties (of a concept of "species"), be they wild or cultivars. This labelling follows the ordinary language used by professionals in seed banks and relates to breeding genetics. However, as seed banks are collections, it is also inspired by the taxonomic concept of *specimen* in natural history collections. While museums materially collect specimens, the conceptual object of the collection is the *type*. It is embodied in specimens that exemplify the defining features of a particular taxon. That said, explanatory tensions occur when seed banks refer to their objects and different purposes of collection. In seed banks, the overall conceptual object of collecting is the gene pool. This refers to intra-species populations, as will emerge in due course.

In what follows, I will mainly take the perspectives of philosophy of science and technology – rather than the perspective of plant ethics. However, the aim of this chapter is to highlight important conceptualizations of what a seed is or can be, and some related issues in biobanking, and this may be useful to plant ethics. To date, plant ethics has had little to say about seed banks; but in the future it may critically refer to these institutions in terms of protecting plants' potentiality and continuity. That, however, would bring on a cluster of epistemological and ontological questions, which I will try to sketch in closing.

Seed banks and biodiversity

By definition, seed banks operate in both past- and future-oriented ways. "A seed bank is an organization or facility used to store and preserve plant seeds, in particular varieties that are rare, have fallen out of commercial use, and/or may have unique desirable genetic characteristics" (Cohen 2014, 1642). This definition shows the ambivalent position in which modern seed banks are situated: between nature conservation and innovative breeding options. Seed banks should prevent gene erosion (i.e. the loss of biodiversity in a genetic sense) while forming an infrastructural alliance of banking and breeding for agro-biotechnology.

The UN Convention on Biodiversity (CBD) has defined the legal framework of conservation (UN 1992). To the disappointment of many environmental thinkers and activists, article 9 of the CBD has facilitated the protection of biodiversity by means of *ex situ* conservation strategies including seed banks, rather than only by protecting wild lands or habitats. At the same time, accompanied by new technologies for rapid genome sequencing, the concept of biodiversity was expanded from the ecology-driven concept of *inter-species diversity* (diversity of different species) to the genetics- driven concept of *intra-species diversity* (diversity within the gene pool of one species, in case of plants of one genus). Philosophers continue to work on the question of what "biodiversity" denotes for different contextual purposes (e.g. Lanzerath and Friele 2014). How that question is answered will have important implications for the conceptualization of "plants" in the context of seed banking – where a plant can be thought of as anything from a "wild plant" to a "plant genetic resource" (PGR).

More than 1,750 seed banks with *ex situ* collections of plant genetic resources for food and agriculture exist worldwide (FAO 2010, 55). The majority of these banks are hosted by industrialized countries. This contradicts the recommendation in Art. 9(b) of the CBD, which stresses that a seed bank should be located near the "origin of genetic resources" (thus effectively more in developing than in industrialized countries). There is a sort of path dependence whereby decisions for plant biodiversity conservation are channelled along the tracks of the imperial politics of the past.[3] Accordingly, low-tech versions of agrobiodiversity conservation, such as "on farm conservation", tend to receive comparatively less attention.

Many other seed banks exist for horticultural plants and for "wild plants" in botanical gardens. The latter are usually regarded as implementing the most natural form of *ex situ* conservation. However, the collections of botanical gardens and major seed banks recently have found points of convergence. For seed banks, the collecting of *crop wild relatives* (CWRs), that is the wild types of domesticated plants, has become a high priority for recovering the full range of the crop's gene pool (Castañeda-Álvarez et al. 2016). This is of interest because the pool might contain so-called "elite events" useful for creating new crop varieties (for public communication, these favourable genetic traits are called "breeding options"). For this reason, old botanical gardens that still host rare CWRs are attractive to seed bankers.

Ethical challenges

The growing importance of large seed banks exerts pressure on ethics in many respects. I can only mention some important questions here. First, the emerging field of *biobank ethics* still exclusively concentrates on human cells, tissue and blood. The highest ethical principle employed is the *informed consent* of the donor or patient (Hoeyer 2008) – which is clearly not applicable to plants. Plants themselves cannot agree on the collecting and storing of their seeds. Nonetheless the principle might be applied to seed banks as donor institutions, and to indigenous people who help academically trained gene scouts find and collect samples of rare plant populations in remote regions.

Second, some philosophers who hold a biocentric view in environmental ethics are challenged by the ethics of seed banking to reconsider their somewhat naïve images of "wilderness" and "wildness" when it comes to plant life. A focus on what is "wild" typically excludes crops and agrobiodiversity from consideration, whereas these are currently the main interests of seed banks.

Third, plant ethics must decide what its unit of concern is, or at least distinguish among the many possibilities outlined in philosophy of biology. Importantly, plant ethics will have to overcome *species essentialism*, purporting that species are natural kinds (on the debates, see Ereshefsky 2017). And then it will have to decide what its units of concern are. Are the units of concern individual plants, genes or gene patterns, plant organisms, plant species, plant populations, or other plant concepts and entities (e.g. types)? And how do seeds relate, ontologically and epistemologically, to these units of concern?[4] For seed banks and collectors, the plant population and a crop's gene pool are the relevant biological categories to consider, which implies a dynamic conceptualization of both "seed" and "plant": what are the ethical implications of that?

Fourth, as pertains to *intra-* and *inter-generational justice*, it matters who has access to the stored PGR and is allowed to share its benefits. This refers not only to nation states or actor groups with different types of knowledge (experts vs. lay people) but fundamentally also to farmers and peasants, who used to be the professionals for seeds. Women, in particular, were traditionally responsible for seed storage – is there a gender dimension to the ethics of seed banking?

As plant ethicists have rightly noted, plants (excluding trees) are still not on the radar of ethics, though nature has been the focus of ethical reflection in academia at least since the 1970s. In addition, while environmental ethicists have been considerably interested in nature conservation strategies, such as the restoration of wild lands, and have dwelt on discussions of naturalness and artificiality (Birnbacher 2014; Elliott 1998), the practices and institutions of *ex situ* plant conservation have been largely overlooked. These have rather been topics of interest to NGOs and concerned farmers (see Flitner 1995; Fowler and Mooney 1990; Juma 1989), and ethnobotanists and anthropologists (Nazarea, Rhoades, and Andrews-Swann 2013). These groups and academics have often questioned the ways in which seed banks have been positioned within the landscape of conservation policy. It has also been suggested that neo-colonial interests might be green-washed by reference to biodiversity conservation goals (e.g. Merson 2000).

One major environmental philosopher, Holmes Rolston, has recently issued the following warning with reference to botanical gardens: "*Ex situ* conservation is always a means, never an end" (Rolston 2004, 37). This statement is controversial, for many actors involved in seed banking do regard this banking as an end in itself. But Rolston's warning is also an important recognition that *ex situ* conservation, including through seed banking, can promote important ends, such as increasing food security and alleviating world hunger (Plucknett and Smith 1987), fighting genetic erosion, and preserving old crop varieties as "common heritage of mankind" (cf. Hammer 2003). The diversity of such ends might generate conflicts: not just within the institutions involved (with regard to the different research policies that they have to adjust to – cf. Karafyllis und Lammers 2017), but also in different understandings of the objects of these institutions' work and in different collecting practices. For example, the pursuit of increased food security will likely result in collecting *genes* as elite events for the known cash crops (wheat, maize, rice and a few others), whereas preserving old crops as "heritage of mankind" would have to focus on *seeds* of non-commercial plants that do not meet present eating habits of the global masses (but maybe only of some groups). Therefore, the concept of continuity also matters with regard to cultural entities and their histories.

In sum, seed banks are valuable in various ways, but their purposes are still under-theorized. This calls for ethical reflection. The challenge is to make ethical norms and values explicit while covering the complexity of the discursive field of seed banks. Such field, as I will highlight throughout the next section, is crucially dependent on technology and its concepts.

Seed banks as "Bestand" for making biofacts, or: Against the wilderness fallacy

Seed banks can be regarded as set of technical conditions for controlling plants' potentiality and continuity in a threefold sense: as stock ("Bestand"), as infrastructure, and as conservation site. It is in the latter context that Rolston (2004) argues against "captive plants" forced to live in botanical gardens rather than in the wild. This position is liable to what I call the "wilderness fallacy", which is arguably inspired by analogy with the case of animals. The "wilderness fallacy" consists in the assumption that the wild and the domesticated or cultivated are neatly divided. It is also part of the fallacy to conflate wilderness with "wildness" (see Kirchhoff and Vicenzotti 2014 for details), which can in fact be exhibited by the many plants that indeed "went wild" and escaped from botanical gardens (e.g. by seed dispersal). From the wilderness perspective, these plants are not wild, but invasive.

More generally, the wilderness fallacy makes it difficult to align environmental ethics with agricultural ethics. In addition, the molecular genetic conception of "wild" – plant populations having more genes (alleles) than their domesticated, hybrid relatives – has not hit the radar of plant ethicists yet. But the structural downgrades and upgrades of plant genomes matter for seed bankers and breeders.

It is noteworthy that seed banks, while being sites of conservation, are also *research infrastructures*. Rolston, like many others, fails to appreciate the mediality (i.e. the enabling or creative potential within the continuous process of mediating) of these specific kinds of infrastructures, how they build stock and facilitate the flow of germplasm. It still remains vague what kind of research they predominantly mediate, however.

Viewed philosophically, the collecting and storing of seeds build *stock* – or, in Heidegger's words (1991, 16), "Bestand", which literally means a "standing reserve". This stock, comparable in some ways to an arsenal, builds the base for technical endeavours. As stock, the entities concerned (in our case seeds) lose their identity and material objectivity (in our case being a plant). On Heidegger's description, "Bestand" is a storage of means rather than an assembly of things or artifacts. These means can serve different ends, and all the more so when these means are living entities. Elsewhere I have called this techno-natural condition of living entities "biofact" (Karafyllis 2006). Biofacts are entities that grow but that no longer grow by themselves. Instead, they are *made* to grow. At the same time, their growth becomes a means to some other end. The living entity does not show this instrumentality, which has to be unveiled through hermeneutical exploration and the historical re-contextualization of episodes of techno-genesis. How does this relate to collecting and storing seeds, which are the oldest agricultural entities configured as "Bestand"?

The making of every plant biofact starts with exerting control on the seed. However, since agriculture has come to be shaped by genetic and biotechnological methods in the late nineteenth century, the methods of seed collecting and storing have radically changed. A worldwide hunt for lowbred crops with rare mutations, at that time called "polymorphisms" (Vavilov 1922), started in the 1920s. Located at the plants' assumed "centers of origins", these rare local varieties, often bred by indigenous people, were and still are regarded as raw material for future breeding options.[5] Consequently, the "Bestand" is constituted through both processes of nature and actions of people. However, from within a genetic perspective that models the plant as a functional aggregate of genes, it became plausible to think that the conservation of seeds is ultimately unnecessary, as long as the relevant genetic material is conserved for synthetic biology. The material might even be reduced to its virtual codes. The stock of a seed bank would then have been transformed into the stock of a data bank.

The object of conservation for a seed bank is the plant's gene pool, and this is embodied by both seeds and datasets denoting variability. Usually, a plant's gene pool refers to the genus and includes the "crop wild relatives". Among these, for example, are the grasses that were the original breeding material for cereals. Later, these grasses became weeds from the perspective of those involved in agriculture. Famous examples for the Americas are the teosintes, part of the evolutionary trajectory of maize in Mexico, the southern USA, and the Andean region. The classification of teosinte varieties can range from a noxious weed to a companion plant of maize (facilitating introgression). This means that the gene pool of a crop is not only a biological category but also a *historical* category, referring to geographical plant origins (commonly known as "centers of origin" or "Vavilov centers").

In addition, the gene pool comprises different regional clusters of crop evolution (i.e. early breeding) and is thus also a *cultural* category, as these clusters track early human settlements and migrations. It is against this backdrop that seed banks present themselves as being in the business of conserving the "common heritage of mankind". If so, they protect (agri)-culture rather than nature; and hence resemble museums.

Object configurations: The seed in seed science and technology

Seeds in the true sense are fertilized ovules, the result of successful pollination in flowering plants and gymnosperms. Each seed consists of a genetically unique embryo, usually accompanied by some food supply and an outer protective covering. If this covering allows desiccation (a precondition for deep freezing), then seed bankers speak of "orthodox" seeds. By contrast, "recalcitrant" seeds are difficult or even impossible to store by conventional methods. This terminology was introduced in 1973 by Eric H. Roberts (Roberts 1973), initiator of the Seed Science Laboratory at the University of Reading, UK. Note that the somewhat new field of "seed science" has created plant-related terminologies that crucially differ from botanical terminologies and treat the seed as an object for long-term storage and building stock. In storing contexts, for example the botanical concepts of seed dormancy and quiescence are transformed into "seed longevity", which, at close look, is a technical category. Therefore, longevity in dry storage is "a poor predictor of persistence in the soil" (Thompson 2000, 217). Building an analogy is limited to hard-seeded species ("orthodox seeds").

Regarding seed banks technology, imagine thousands of jars used as seed containers for orthodox seeds, lined up on shelves in large cooling chambers. By means of cooling systems, banked seeds are kept chilled and remain viable for decades. The "natural" time span between the two abstract plant entities, seed and seedling, is the object of technical prolongation. However, as outlined in the beginning of this chapter, there is no such natural time span, at least not a distinct one. Moreover, plant science knows many varieties of seed dormancy (physical, chemical, morphological, etc.) and quiescence, most of them are still poorly understood. For technical modelling of artificial dormancy, seed banks require protocols for each plant species. Subsequently, by resisting standardization efforts, seeds also "walk on the wild side" within the seed bank. On average, seed longevity under modern storage conditions lasts about 30 years; much shorter for the fatty seeds of oil plants, much longer for peas.

If cryoconservation techniques for tissue storage in liquid nitrogen are applied (e.g. ultra-rapid freezing), then the germplasm may even be conserved for centuries. These methods are the culmination of a whole century of research on plant tissue culture (Gautheret 2003) and *in vitro* methods. Seed banks use these methods for storing the many plants that reproduce mainly vegetatively by bulbs, shoots and tubers, technically speaking by "clones". These plants have been bred to do so or have been selectively proliferated by farmers in that way in order to keep the cultivar consistently identical (as in the case of potato and banana cultivars). It is noteworthy, that plants have different reproductive systems. Many can even switch between them. This might be interpreted as another form of a plant's wildness, which is object of standardisation within the seed bank.

In sum, storing seeds and germplasm occurs within "cryogenic time regimes" (Friedrich 2016). They shift the limits not only between life and death but also between the natural, the agricultural and the technical modes of being. The many different life forms and varieties of plant life remain challenging with regard to technical standards. The related methods pose the ontological question of what a seed is, and can be.

Aristotle and Theophrastus: What is a seed, and how can it matter?

Let us begin by stressing that the plant entities to be banked as reproductive units do not necessarily have to be seeds. Scientists in seed banking prefer to speak of "germplasm" rather than "seed" as the lowest common denominator of reproducible plant life. The totipotence of the plant cell makes a considerable amount possible in biobanking and biotechnology (Börner 2006; Engelmann 2011). Here, however, I will still generalize all possible reproductive units of plants as seeds, as the seed was, and in many cases still is, the boundary object *and* the material object for establishing seed banks. Doing so also makes it possible to relate seed banks to the history of agriculture and, furthermore, to Aristotle's and Theophrastus' philosophical understanding of "seed".

We owe what is perhaps the first description of a seed bank to the Aristotelian scholar Theophrastus. In the *Enquiry into Plants* (Latin: *Historia Plantarum*, HP), Theophrastus writes of seeds "which keep or do not keep well" under storage conditions:

> The seeds have not all the same capacity for germination and keeping well. Some germinate and mature very quickly, and keep excellently, as Italian millet and millet. Some germinate well, but soon rot, as beans, and especially those that are "cookable"; so do tare and calavance; also barley perishes sooner than wheat.
>
> (Theophrastus 1926, HP VIII, 11, 1)

It is noteworthy that Theophrastus focuses on the two criteria that are at the core of the technical modelling of seeds: *longevity* and *viability* (see Dickie et al. 1990). When a seed germinates, this is the definitive evidence that it remained "alive" during quiescence and is thus capable of resuming growth –, that is that it is viable. Not only in Theophrastus's botany but also in today's seed science, the seed itself has a life, although this life exists only in latency. The minimum criterion of life is metabolic activity. A seed in the cooling chambers still has metabolic activity – and this is in fact a massive problem for maintaining the seeds' "genetic integrity" by technological means (cf. Börner 2006).

We can say that *a banked seed is a plant persisting in survival mode*. Persistence over time, in this context, then should be addressed with the metaphysical expression "perdurance" (as opposed to "endurance"), implying that an object consists of one or more temporal parts at every moment it exists. This object does not necessarily have to be an individual; an insight, which allows for relating the individual seed to the *continuity* of its population, and not only to the becoming of an individual plant. In sum, the natural seed is not an object of endurance but perdurance.

As Aristotle (1953) informs us in the *Generation of Animals* (GA), a "plant, in its essence, has no function or activity to perform other than the production of its seed" (GA I 731a 24). On this view, more than all other plant characteristics, proliferation by seeds marks plant life. The reason for this strong statement is that Aristotle's idea of plants is the model for his concept of the *vegetative soul*, which enables growth, nutrition and reproduction in all living beings. To separate in thought the plant from its reproductive potential means, ontologically, to disable the very essence of a living being, that is its process of becoming – which relates to the overall continuity of life. The seed itself (Greek: *spérma*) is also the framing concept for the male semen in animals and humans, and is denoted by the same word. However, Aristotle reminds us, plants "produce not semen but a fetation – what we call then 'seeds' [*spérmata*]". In plants, "the female is not separate from the male, their faculties are mingled together" (GA I 731a 1–7; cf. GA II 741a 5).

That Aristotle does not conceptualize plants in general as asexual shows in his statements on trees that proliferate by shoots or buds: "some living things produce no semen at all", such as "willows and poplars". To produce no semen or seeds is a sign of "weakness" and inferiority (GA I 726a). Note that high-tech seed banking favours vegetative forms of reproduction (which Aristotle exemplified by willows and poplars). Plant's *sexuality*, materialized in the form of seeds,[6] is understood as a risk factor for collecting and sorting practices that refer to the distinguishable plant type as the central ordering (and breeding) category. Seeded propagation, however, results in slightly different versions of the parent. As a consequence, sexuality questions the offspring's *identity*, and so does the seed. Yet sexuality is an advantage for the continuity of the population under natural conditions. It can be acknowledged as the overall plant potential to "walk on the wild side" of life.

From the point of view of natural history, seed banks' new preference for clonal tissue not only turns the natural order upside down but is also a backlash in evolution. What both Aristotle and Darwin regarded as a lower stage of plant life (vegetative reproduction) becomes central once technology is in the picture, as technology aims at a standardized object. In this sense, high-tech seed banks can be viewed as transforming seeds into a biotechnically formable material.

Another point in Aristotle's interpretation of seeds matters for plant ethics. At the beginning of its "action", the vegetative soul cannot be separated from its initializing medium, that is here, the soil. Like a small technician (Aristotle indeed uses the Greek *téchne*), the vegetative soul within the seed organizes its growth milieu by movement, as though having the tools of heat and cold ready at hand (GA II 741b 25ff.). The potentiality is within and – at the outset of germination – around the seed, in its medium. Together, the seed and its medium are the substance (Greek: *ousía*) of becoming. The seed in its perdurance also embodies both aspects of *arché* – the origin and the beginning – while at the same time being a body (or bodily object) without having one. By this construction, the seed denotes not only potentiality but also *continuity* of life (resembled in the modern term "germplasm").

Potentiality, continuity, perdurance

As we have seen, conserving plants' potentiality while interrupting continuity is the central methodological goal of seed banking. Reference to potentiality clearly occurs at the level of genetics, as well as on at least four other levels:

1. the potential of a seed (in the broad sense of germplasm) to become a plant, which refers to ontogenesis;
2. the potential of a "wild plant" to become a crop by means of breeding and genetic engineering, including the utilization of the gene pool, which refers to biotechnology's innovation potential;

3. the potential of the land to become an acre or a garden by means of agri- and horticulture, which refers to the potential of culture, including cultural techniques (German: "Kulturtechniken");
4. the genetic potential of CWRs as encompassing breeding options for future highbred crops, which refers to potentials on the phylogenetic level (reconsider the "centres of origins") and thus plant evolution together with socio-cultural evolution of mankind.

Consider now the peculiar notion of "plant freedom" that Rolston (2004) appeals to with his expression "captive plants". He only mentions plants in gardens and in pots. While plant freedom is a questionable category for plant ethics when referring to the kept animal as the boundary object, it might be a fruitful criterion for evaluating seed practices. Here it is important to understand "freedom" and "persistence" not in a spatial but in a spatiotemporal sense. "Plant population" then means a *continuant* that consists of different temporal parts and counts as unit of perdurance. One of these temporal parts is "the seed", which (as a singulare tantum) likewise encompasses different temporal parts within a soil seed bank, that is individual seeds and their different time-spans of *enduring* unfavourable conditions. Therefore, only the *individual* seed can be regarded as object of endurance, while the population it signifies is object of perdurance. This insight highlights that seed banks can aggregate individual endurances of seeds (in one accession) but are not able to collect and store the perdurance of the population as a metaphysical surplus, which would guarantee for the population's continuity. Instead, seed banks *transform the intertwined modes of potentiality and continuity into a sum of possibilities*, which appears to be a homogenous conservation unit on the meta-level.

The strength of this philosophical approach is to emphasize that plants in a seed bank have indeed lost a certain freedom: the freedom to *become* a plant – concerning the naturally combined modes of *potentiality* (what to become) and *continuity* (when to become) – is constrained by storage technologies. They turn plant's potentials into options.

First, the collected seed samples are decontextualized from their origins, foremost in the phylogenetic sense of the word. This compromises Aristotle's concept of *arché* (see above). Subsequently, seed banks primarily collect possible beginnings, not potential origins, which is disguised in the phrase of "collecting breeding options". Even if they try to complete the gene pools of crops by means of collecting CWRs, they regard these origins as secondary to beginnings. Therefore, seed banks tend to devalue plant-associated values from the past and overvalue estimated assets of the future.

Second, seed banks *synchronize* the genuinely diachronic germination potentials of seeds belonging to one population. At close look, they increase the possibility of the (assumingly) 1,000 *individual* seeds to *endure* within a jar, while compromising the *perdurance* of "its kind" on the level of population development. Therefore, if ethicists want to speak of "captive plants", then it is not only the hindered potential of becoming the next "generation" of a plant that they might want to emphasize. What really matters is that seed banks have to manage seeds as enduring materials and not as perduring substances, in which different time parts are interwoven (e.g. also the times of the seasons). Therefore, seed banks cannot collect the interactions of seeds with their media (soil, pollinators, etc.), which would account for the full potentiality of seeds. As a consequence, a seed in a seed bank shows that it "walks on the wild side" by being recalcitrant to storing conditions, or even – paradoxically – by its death.

Summary and outlook for plant ethics

This chapter has shown the argumentative challenges for plant ethicists who, in the future, might turn to evaluate seed banks. While I honestly doubt that a plant's "right to flourish" is a decisive ethical argument against making and storing biofacts – as many plants do not have blossoms and as many crops have been prevented from flourishing by our ancestors long before the age of biotechnology – I believe that plant ethics can offer valuable arguments for a plant population's *interest in continuing*. Whatever else is true, this is the interest that seed banks protect, even if only in a reduced manner and with reference to only few species. Seed banks turn seeds into stock and decontextualize plant life from its media. These institutions do so because anthropogenic changes in the ecological world, such as climate change, have placed plant populations of not only cultivars at risk. However, conserving plant evolution by means of a time standstill in the bank is risky, too. To say the least, this contradicts the scientific concept of co-evolution: of plants and pollinators, pathogens, and others, not to forget humans. Therefore, it remains important to fight the idea that seed banks are substitutes for conserving plants in natural habitats (i.e. *in situ*) or agricultural ecosystems (Nazarea, Rhoades, and Andrews-Swann 2013).

While nobody can fully guarantee a sound functioning of long-term stored seeds when transferred into outdoor environments again, it is still better to collect than to neglect seeds of rare plants. However, as long as the major national seed banks are not explicitly serving conservation issues to a larger extent (for instance by contributing to the repatriation of wild plants and "on farm conservation" of rare landraces), they remain exposed to the critique that they protect biodiversity mainly to breed uniformity (Saraiva 2013). Hence the seed would matter only as genetic reservoir or stock, not as future plant.

Technology – in this case seed banking technology – continues to transform botany and agriculture not only with regards to their methods but also with regards to their *concepts*. The results are epistemic-moral hybrids, which should not be taken as a given by ethicists but, rather, be regarded as objects that need deconstruction. Agrobiotechnology is generally more interested in clonal tissue rather than in seeds. This has influenced the activities and choices not only of seed banks but also of biobanks more generally. Therefore, it makes sense to analyze the different "orders of things", to use Foucault's words, and how "living things" in biobanks are transformed into stocks, options and assets, before moving onto the ethical evaluation of plants and seeds.

In closing, I would like to summarize my recommendations for future plant ethics as follows:

- explicitly include cultivars in the concept of "plant" (i.e. plants as biofacts);
- take into account the gene pool of a plant, and hence the plant population, instead of the plant species or individual only;
- broaden the understanding of a plant's "wildness" and "wilderness" to the insight that crop wild relatives materialize genetic preconditions for breeding, including next-generation genetic engineering (e.g. CRISPR/Cas) and synthetic biology;
- question the assumption that wild plant conservation and biodiversity protection oppose biotechnological use of the plants concerned;
- conduct basic research on the ontology of seeds, including their technically shaped asexual equivalents such as clonal tissue.

At last, we might remember another famous song of Lou Reed: "Perfect Day" (1972). It closes with the repeated phrase "You're going to reap just what you sow".

Acknowledgments

This research results from the research project "The Language of Biofacts: Materiality and Semantics of High-Tech Plants" (2015–2017) and was supported by the *German Federal Ministry of Education and Research* (BMBF). I thank the many interview partners from seed banks who provided expertise that greatly assisted the research, although they may not agree with all of the interpretations of this paper.

Notes

1. For a short historical overview, see Scarascia-Mugnozza and Perrino (2002).
2. In collection theory, "accession" means the permanent record of a received object, that is an object that has been accepted as registered part of the collection. In seed banks, the accession consists of various seed samples of one kind (of a cultivar or a wild species), which is conceptualized as homogenous after a thorough evaluation process.
3. On the history of seed banks see Flitner (1995), Heim (2008) and Karafyllis and Lammers (2017) for Germany; Saraiva (2013) for Europe and Juma (1989) for the Americas.
4. For example, if "species" means interbreeding lineage (cf. Ereshefsky 2017), then this depends on sexuality. As a consequence, seed producing plants have an ontological status within this explanatory model of "species", whereas cryptogams have not.
5. The systematic process of utilizing a crop's gene pool was highlighted by the father of modern seed banking Nikolaj I. Vavilov (cf. Vavilov 1922).
6. Here, I cannot not comment on the vegetative forms of seed production (apomixis or agamospermy).

References

Aristotle. 1953. *Generation of Animals*, transl. by A. L. Peck, revised ed. Cambridge, MA; London: Harvard University Press.

Birnbacher, D. 2014. *Naturalness*. Lanham, MD: University Press of America.

Börner, A. 2006. Preservation of plant genetic resources in the biotechnology era. *Biotechnology Journal* 1, no. 12: 1393–404.

Castañeda-Álvarez, N. P., Khoury, C. K., Achicanoy, H. A., Bernau, V., Dempewolf, H., Eastwood, R. J., Guarino, L. et al. 2016. Global conservation priorities for crop wild relatives. *Nature Plants*, no. 2: 16–22.

Cohen, S. 2014. Seed banking, seed saving, and cultivating local varieties. Pp. 1642–48 in *Encyclopedia of food and agricultural ethics*, edited by P. B. Thompson and D. M. Kaplan. Dordrecht: Springer Netherlands.

Dickie, J. B., Ellis, R. H., Kraak, H. L., Ryder, K. and Tompsett, P. 1990. Temperature and seed storage longevity. *Annals of Botany* 65, no. 2: 197–204.

Elliot, R. 1998. *Faking nature: The ethics of environmental restoration*. London: Routledge.

Engelmann, F. 2011. Use of biotechnologies for the conservation of plant biodiversity. *In Vitro Cellular & Developmental Biology – Plant* 47, no. 1: 5–16.

Ereshefsky, M. 2017. Species. In *The Stanford encyclopedia of philosophy* (Fall 2017 ed.), edited by Edward N. Zalta. Accessed January 7, 2018 from https://plato.stanford.edu/archives/fall2017/entries/species/.

FAO. 2010. *The second report on the state of the world's plant genetic resources for food and agriculture*. Rome: FAO. Accessed December 5, 2017 from www.fao.org/docrep/013/i1500e/i1500e00.htm; http://www.fao.org/docrep/013/i1500e/i1500e.pdf.

Flitner, M. 1995. *Sammler, Räuber und Gelehrte: Die politischen Interessen an pflanzengenetischen Ressourcen 1895–1995*. Frankfurt am Main: Campus.

Fowler, C. and Mooney, P. R. 1990. *Shattering: Food, politics, and the loss of genetic diversity*. Tucson, AZ: University of Arizona Press.

Friedrich, A. 2016. Die Vergänglichkeit überlisten. Leben und Tod in kryogenen Zeitregimen. Pp. 35–57 In *Jahrbuch Technikphilosophie 2016*, edited by G. Gamm, P. Gehring, C. Hubig, A. Kaminski and A. Nordmann. Zürich, Berlin: Diaphanes.

Gautheret, R. J. 2003. Plant tissue culture – The History. Pp. 105–13 in *Plant tissue culture: 100 Years since Gottlieb Haberlandt*, edited by M. Laimer and W. Rücker. New York: Springer Vienna.

Hammer, K. 2003. A paradigm shift in the discipline of plant genetic resources. *Genetic Resources and Crop Evolution* 50, no. 1: 3–10.

Heidegger, M. 1991. *Die Technik und die Kehre* (8th ed.). Pfullingen: Neske.

Heim, S. 2008. *Plant breeding and agrarian research in Kaiser-Wilhelm-Institutes 1933–1945: Calories, caoutchouc, careers*. Boston: Springer.

Hoeyer, K. 2008. The ethics of research biobanking: A critical review of the literature. *Biotechnology & Genetic Engineering Reviews* 25, no. 1 (January): 429–52.

Juma, C. 1989. *The gene hunters: Biotechnology and the scramble for seeds*. Princeton, NJ: Princeton University Press.

Karafyllis, N. C. 2006. Biofakte – Grundlagen, Probleme, Perspektiven. *Erwägen Wissen Ethik* 17, no. 4: 547–58.

Karafyllis, N. C. (ed.). 2018. *Theorien der Lebendsammlung. Pflanzen, Mikroben und Tiere als Biofakte in Genbanken*. Freiburg: Alber.

Karafyllis, N. C. and Lammers, U. 2017. Big Data in kleinen Dosen. Die westdeutsche Genbank für Kulturpflanzen 'Braunschweig Genetic Resources Collection' (1970–2006) und ihre Biofakte. *Technikgeschichte* 84, no. 2: 163–200.

Kirchhoff, T. and Vicenzotti, V. 2014. A historical and systematic survey of European perceptions of wilderness. *Environmental Values* 23, no. 4: 443–64.

Kohler, R. E. 2006. *All creatures: Naturalists, collectors, and biodiversity, 1850–1950*. Princeton, NJ: Princeton University Press.

Lanzerath, D. and Friele, M. (eds.). 2014. *Concepts and values in biodiversity*. London: Routledge.

Merson, J. 2000. Bio-prospecting or bio-piracy: Intellectual property rights and biodiversity in a colonial and postcolonial context. *Osiris* 15, no. 1: 282–96.

Nazarea, V. D., Rhoades, R. E. and Andrews-Swann J. (eds.). 2013. *Seeds of resistance, seeds of hope: Place and agency in the conservation of biodiversity*. Tucson, AZ: University of Arizona Press.

OECD. 2001. *Biological resource centres: Underpinning the future of life sciences and biotechnology*. Paris: OECD.

Pistorius, R. 1997. *Scientists, plants and politics: A history of the plant genetic resources movement*. Rome: IPGRI.

Plucknett, D. L. and Smith, N. J. H. 1987. *Gene banks and the world's food*. Princeton, NJ: Princeton University Press.

Roberts, E. H. 1973. Predicting the storage life of seeds. *Seed Science and Technology* 1: 499–514.

Rolston, H., III. 2004. In situ and ex situ conservation: Philosophical and ethical concerns. Pp. 21–39 in *Ex situ plant conservation: supporting species survival in the wild*, edited by E.O. Guerrant Jr., K. Haven and M. Maunders. Washington: Island Press.

Saraiva, T. 2013. Breeding Europe: Crop diversity, gene banks, and commoners. Pp. 185–212 in *Cosmopolitan commons: Sharing resources and risks across borders*, edited by N. Disco and E. Kranakis. Cambridge, MA: MIT Press.

Scarascia-Mugnozza, G. and P. Perrino. 2002. The history of ex situ conservation and use of plant genetic resources. Pp. 1–23 in *Managing plant genetic diversity*, edited by J. M. Engels, V. Ramanatha Rao, A. H. D. Brown and M. T. Jackson. Wallingford, UK: CABI.

Theophrastus. 1926. *Enquiry into plants, and minor works on odours and weather signs*, transl. by A. Hort, Vol. II (reprint). Cambridge, MA; London: Harvard University Press.

Thompson, K. 2000. The functional ecology of soil seed banks. Pp. 215–35 in *Seeds. The ecology of regeneration in plant communities*, 2nd ed., edited by M. Fenner. Wallingford, UK: CABI.

UN. 1992. *Convention on biological diversity*, Rio de Janeiro: UN. Accessed January 13, 2018 from https://treaties.un.org/doc/Treaties/1992/06/19920605%2008-44%20PM/Ch_XXVII_08p.pdf.

Vavilov, N. I. 1922. The law of homologous series in variation. *Journal of Genetics* 12, no. 1: 47–89.

17 CRISPR/Cas in crop breeding

Why ethics still matter

Frauke Pirscher

Introduction

CRISPR/Cas[1] is one of the newly developed genome editing techniques viewed as revolutionary to crop breeding. It allows a precise modification of genes by adding, cutting or suppressing certain gene sequences of the DNA. Compared to earlier breeding techniques for genetic modification (GM), this system is considered to be easier to apply, more precise, much quicker, and therefore cheaper (Baker 2014). CRISPR/Cas is expected to have great innovative potential in agriculture by speeding up breeding, increasing yields and allowing plant production to occur under less favorable conditions. It has already been successfully applied in the breeding of different agricultural plants such as soybean (Jacobs et al. 2015), maize (Svitashev et al. 2015), tobacco, sorghum, rice (Woo et al. 2015), and tomato (Brooks et al. 2014).

Although CRISPR/Cas generally allows for transgenesis (the transfer of DNA sequences across species), its current application in agricultural biotechnology remains mainly cisgenic (within species boundaries). CRISPR/Cas-based cisgenesis allows for creating new products that could in principle also be the outcomes of natural evolution or conventional breeding techniques. It also offers the possibility of "reverse breeding", that is the re-introduction of properties that were lost in the course of breeding (Palmgren et al. 2015). When using common tests the genetic modification of DNA could not be traced in the new products up to now (Ahmed 2002; Araki and Ishii 2015). Thus, the cisgenic products cannot be identified as genetically modified organisms (GMOs). Nucleotides (DNA and RNA) can be synthesized so that they are molecular equivalent to the native temples. Therefore, they are so-called nature-identical GMOs.[2]

These new options for breeding have shifted the societal debate on genetic engineering in a new direction, opening a variety of controversies. One of these is whether cisgenic plants should be viewed as genetically modified at all, considering that cisgenic plants can be entirely indistinguishable from plants that are the outcomes of traditional breeding techniques. The answer that is given to this question has implications regarding the nature of the regulations that may be applied to these products, and even the need for applying regulation of any kind. This, in turn, may significantly influence the profitability of the innovation. A second controversy concerns the permissibility of reverse breeding techniques in organic farming. Within both debates, the concept of naturalness plays a key role and is used as an argument both by proponents and opponents of CRISPR/Cas.

Against this background, the aim of this chapter is threefold. First, reconstruct the societal debate on the application of CRISPR/Cas and its regulation; second, analyze the different understandings of naturalness that lead to different judgments about the moral acceptability of the technology; and third, argue that natural identity is not a sufficient argument to reject all ethical concerns that can be moved to CRSIPS/Cas.

Reconstruction of the societal debate

A key point within the debate on the use of CRISPR/Cas in plant breeding is whether cisgenic plants should be viewed as GMOs or not. Current EU Law defines GMOs "an organism, with the exception of human beings, in which the genetic material has been altered in a way that does not occur naturally by mating and/or natural recombination" (European Parliament 2001, 4).

EU jurisprudence on GMOs was developed at a time when GMOs meant introducing transgenic DNA into a genome. Thus, the regulation does not distinguish between transgenic and cisgenic plants, and defines GMOs by looking at the process of creation and not its outcomes. If a technique is defined as GM, it has to undergo a special risk assessment that substantially increases the cost and time needed for approval. At the same time, EU law allows for exceptions from this procedure in cases of mutagenesis and cell fusion (European Parliament 2001, 5, 18). If cisgenic plants are viewed as GMOs, then the advantage of rapid breeding success can be neutralized by the time that is needed for the approval procedure. This has negative consequences for the profitability of the innovation. Furthermore, products under EU regulation have to be labelled as GMOs, which might have a negative consequence for consumer acceptance. Thus, the categorization of CRISPR/Cas as a genetic modification is both a societal question of risk management and an economic question of profitability for breeding companies.

Unsurprisingly, the leading breeding companies call for a product-oriented regulation that would not define cisgenic plants as GMOs. Their key argument relies on the natural identity of the modified and traditional plant (KWS 2017). This argument has also been made by scientists (Araki and Ishii 2015; Huang et al. 2016; Schouten, Krens, and Jacobsen 2006), and it holds that cisgenic plants obtained through CRISPS/Cas techniques are to be assimilated to traditionally bred plants because the transferred genes originate from the same gene pool. According to this argument, no changes in fitness occur that could not also occur through either traditional breeding techniques or natural gene flow. Hence, cisgenesis carries no distinctive risks (e.g. regarding potential effects on non-target organisms or soil ecosystems, toxicity or possible allergy risks emanating from GM food or feed) other than those that are also incurred with traditional breeding techniques. Therefore, leading breeding companies call for an approval procedure of CRISPS/Cas products like that used for products of traditional breeding (Carroll et al. 2016; Huang et al. 2016).

In contrast, critics argue that although the product might be identical to those developed by nature, the process is highly artificial. All GM techniques directly interfere at the level of the genome by inserting material that was produced outside the cells. This degree of intrusion into the cell is viewed as a risk because the consequences for the plant are not sufficiently known. Critical biologists, such as Then (2016) and Steinbrecher (2015), argue that CRISPR/Cas is indeed more precise than other GM techniques, but that it is still not free of errors. They note that off-target effects might occur because of insufficient specificity of the nuclease, which causes the DNA to be cut in several regions of the genome instead of the one intended. Therefore, critics call for a regulation of the process, as is currently the case under the EU regulation, to guarantee the traceability of the genetic modification and to protect consumers' right to know which food products are produced from genome-edited plants. From the critics' perspective, the safety of the technique cannot be guaranteed reliably, and therefore a regulation of the modification process is urgently needed.

Similar divergences become apparent in the discussion about the acceptability of CRISPR/Cas for organic breeding. Organic farmers have different breeding objectives than regular farmers because of their general rejection of synthetic fertilizer and pesticides. Therefore, they need cultivars adapted to low-input growing conditions, with a high resistance to pest and disease (Nuijten, Messmer and Lammerts van Bueren 2017).

The introduction of genes from older varieties that are thus adapted and resistant offers a possibility to better satisfy the specific requirements of organic farming. The application of the CRISPR/Cas technique in breeding would speed up their breeding successes tremendously. Yet when Urs Niggli, head of the Swiss Research Institute of Organic agriculture "Fibl" (Forschungsinstitut für biologischen Landbau), proposed CRISPR/Cas as a possible option in organic breeding (Maurin 2016), the majority of members of the organic farmers' association distanced themselves from his position (Bio Suisse 2016; Bioland 2016). For them, the CRISPR/Cas technique is a genetic modification and therefore not compatible with the principles of organic breeding and farming. Therefore, IFOAM EU, part of the International Federation of Organic Agricultural Movements, calls for additional risk assessment, and for labelling CRISPS/Cas products under GMO regulation in order to distinguish non-GM production and increase transparency and freedom of choice for farmers and consumers (IFOAM 2015).

In the scientific literature, there are opposing positions regarding the compatibility of organic farming with new breeding techniques such as CRISPS/Cas. Generally, organic farming interprets agriculture as a process that includes agro-ecological, socioeconomic, and ethical principles. IFOAM has defined four core principles to reflect this holistic understanding of agriculture: health, ecology, fairness and care (IFOAM 2014). Opponents of CRISPR/Cas have argued that this new way of breeding conflicts with the principle of health, if that principle is understood as supporting the wholeness and integrity of living systems. From this biocentric perspective, any alteration of DNA violates the integrity of the genome as part of a living entity (Nuijten et al. 2016; van Bueren et al. 2003). In contrast to this, proponents note that reverse breeding can reduce the need for chemicals and is therefore not only compatible with organic farming but also urgently needed. It *promotes* the health of the plant, and is therefore compatible with the principle of health: thanks to it the plant can better cope with adverse environmental conditions. Here, reverse breeding is viewed as a return to nature rather than a further departure from it (Andersen et al. 2015; Palmgren et al. 2015).

Breeding between biomimicry and biofact

Different opinions about the need and acceptability of human interference with nature can lead to opposing positions regarding the treatment of CRISPR/Cas modified plants. It must be remembered that breeding is by definition human interference with natural processes, as the ultimate aim of breeding is to change plants to suit human needs. Humans have done this since the Neolithic. At the same time breeding, and agricultural production more generally, are not entirely artificial, as they also crucially depend on natural systems at the molecular as well as higher levels. No clear-cut boundaries between nature and technology seem to exist, in agriculture. Our explorations of nature up to the molecular level, and decoding and modifying of natural processes, are further blurring the distinction.

At first glance, the current public debate focuses on the legal question of regulation and labeling. But when looking at the arguments of both proponents and opponents of process-targeted regulation, the question of nature or naturalness is clearly central. In this context, the notion of naturalness is used not only for describing different degrees of human interference with plants but also as a moral category (Van Haperen, Gremmen and Jacobs 2012). However, the classification of processes as "natural" or "unnatural" is notoriously ambiguous, and this is in evidence in the opposing positions that those calling for or rejecting the regulation of cisgenic crops take.

The two groups seem to employ different understandings of naturalness. In her article "*Dimensions of Naturalness*", Siipi (2008) classifies the different understandings of naturalness or unnaturalness at play in public and scientific debates in bioethics. These categories are helpful to understand how it can be possible to argue both for and against applying CRISPR/Cas by invoking naturalness.

The two understandings of naturalness that are important for the CRISPR/Cas debate are what Siipi (2008) calls history-based and property-based naturalness. While history-based naturalness emphasizes the origin of an entity, or the kind of modification it has gone through, property-based naturalness looks at the current properties of an entity disregarding the manner in which they appeared. For history-based naturalness, the occurrence of human involvement is the criterion for judging an entity as natural or unnatural (human interference itself is viewed as unnatural). On some views, an entity can only be either natural or unnatural. On other views, different degrees of naturalness are possible depending on the qualities of the human interference, such as intensity or duration. Property-based naturalness, on the other hand, ignores processes and judged naturalness by comparison with models.

With regard to CRISPR/Cas, a "historically natural entity", that is a wild plant with its original molecular composition, is used as the reference model.[3] Proponents of CRISPR/Cas typically have a property-based understanding of naturalness. By indicating that the molecular structure of synthesized genes can be identical with that of natural genes, and by arguing that reverse breeding leads to a back-to-nature crop (Palmgren et al. 2015), they clearly interpret naturalness as property-based similarity to wild entities employed as reference models. Hence new breeds that exhibit similar properties as the wild ones are viewed as natural. The more similarities exist between a new crop and wild historic plants, the more natural the new crop is. Furthermore, CRSIPR/Cas can be viewed as natural because the process of gene transfer in the lab is similar to a biological mechanism (Belhaj et al. 2015). Genome editing via CRISPR/Cas applied by the plant breeder is also, at the same time, a mechanism naturally occurring in bacterial cells: the CRISPR sequences provide bacteria with an immune defense against viral attacks by introducing foreign viral gene sequences into a bacterial DNA (Belhaj et al. 2015). On this showing, CRISPR/Cas is a discovery of a natural mechanism, and as a new breeding technique, it is only mimicking nature.

Understanding new breeding technologies as biomimicry is typical of what Zwart (2009) calls a "techno-science perspective on nature" where nature is viewed as a great source of products and processes and as an inspiring source of knowledge to be discovered, explored and imitated. Nature provides the scientist with the tools and techniques he or she can use to develop new products and procedures to solve human problems (Blok and Gremmen 2016), optimize natural processes to human needs and finally "[. . .] design the ideal plant type" (Koornneef and Stam 2001, 159).

Categorizing a product or a process as natural serves at the same time as a moral relief: according to this property-based understanding of naturalness, no moral concerns arise as long as the new product could at least theoretically come into existence without human interference (Weigel 2017). From this perspective, the product is morally acceptable because it is natural.

Opponents of CRISPR/Cas, on the other hand, typically have a history-based view of naturalness, and focus on the process of human interference with nature and not just (the properties of) its outcomes. On this view, plants modified by CRISPR/Cas techniques are more unnatural than traditionally bred plants because CRISPR/Cas plants have undergone greater human-caused changes than traditionally-bred plants. This is true even if the genotype of the plants is similar. Here, it is the processes as such that raise moral concerns. The demand to regulate cisgenic modified plants under EU GM schemes shows that the key concern is not the transition of species boundaries and the unnaturalness of the outcome, but the unnaturalness of the process of human interference with the cell. On this view, although CRISPR/Cas may in principle occur naturally, the fact that it is now carried out by humans raises moral concerns. This is part of a general criticism of technologies that infringe on natural processes.

Opponents of CRISPR/Cas view these new plants as one step closer to "biofacts" (Karafyllis 2003), where more and more key aspects of nature and naturalness are lost. The term biofact expresses the concern that nature is no longer the non-human – in the Aristotelean understanding of that which just comes into being – but is now shaped and designed by technology and thus becomes an artificially created entity (Aristotle 1980). For opponents of CRISPS/Cas, the fundamental question still lingers of whether the acquisition of natural processes for human purpose is legitimate at all. Molecular biology allows what Lee (2003) calls an ontological transformation of living organism,s from natural beings into "biotic artefacts". Nature changes from something that grows and develops to something that is made (Dabrock 2009). For those who oppose it, this technology is not a step closer to more naturalness but a step closer to achieving control of nature and its processes. On these views, the use of CRISPS/Cas remains a highly controversial and urgent moral question.

Conclusions

The natural identity of CRISPS/Cas-modified plants and conventionally bred plants is an observed empirical fact. The molecular composition of the plants can be the same. Proponents of a deregulated application of CRISPR/Cas argue that because of this natural identity "there is no scientific or other logical reason to single out the "process" of the genome editing for onerous regulation" (Carroll et al. 2016, 479). However, the conclusion connecting the empirical evidence of natural identity to calls for a certain kind of regulation and risk management (or omission thereof) seems based also on unspoken value-judgments that may be well disputed in the context of societal debate. What has become apparent above while specifying the various meanings of "natural" or "naturalness" is that the term itself is ambiguous and thus open to interpretation.

Comparing conventionally bred plants with genetically modified plants on the basis of their molecular composition, and concluding that molecular equivalence leads to equivalent risks, is only one way of deciding of the acceptability of these new technologies. In contrast, one might compare plants on the basis of their phenotypes, or their interactions with the environment in a more holistic vein; or one might raise general deontological objections against the disturbance of the genome – as the debate on the application of CRISPR/Cas in organic farming shows. Thus, differences in the acceptance of the new technology are not solely caused by a lack of knowledge or irrationality of lay consumers and citizens, as often suggested by "techno-science" enthusiasts, but are traceable to different value-judgments about the extent to and the ways in which humans should interfere, and be allowed to interfere, with nature. Concluding from natural identity that the risk of the new technology is generally acceptable is to take a moral stance, not to enunciate a matter of fact. Natural scientists in biotechnology need to be aware that ethics is intrinsic, not extrinsic to their task (Bruce 2002). That means that the broader public has to be not only informed about, but also involved in, decision-making on new technological developments such as CRISPS/Cas. A societal discourse on the guiding values regarding the development and application of any new technology should be an integral part of natural science, beginning in the research design phase.

Notes

1. Clustered regulatory interspaced short palindromic repeats, Cas is a CRISPR-associated protein.
2. Kim and Kim (2016) question this assumption by presenting some cases where cells transfected with CAS9 and guide RNA did contain small insertions of foreign DNA at off-target sides.
3. Siipi (2008) further identifies "normality", understood as a statistical or functional concept, and "human nature", as two other possible reference models.

References

Ahmed, F. E. 2002. Detection of genetically modified organisms in foods. *Trends Biotechnology* 20, no. 5: 215–23.

Andersen, M. M., Landes, X., Xiang, W., Anyshchenko, A., Falhof, J., Osterberg, J. T., Olsen, L. I. et al. 2015. Feasibility of new breeding techniques for organic farming. *Trends in Plant Science* 20, no. 7: 426–34.

Araki, M. and Ishii, T. 2015. Towards social acceptance of plant breeding by genome editing. *Trends in Plant Science* 20, no. 3: 145–9.

Aristotle 1980. *Physics. The Loeb Classical Library*. Cambridge, MA: Harvard University Press.

Baker, M. 2014. Gene editing at CRISPR speed. *Nature Biotechnology* 32, no. 4: 309–12.

Belhaj, K., Chaparro-Garcia, A. Kamoun, S., Patron, N. J. and Nekrasov, V. 2015. Editing plant genomes with CRISPR/Cas9. *Current Opinion in Biotechnology* 32: 76–84. https://doi.org/10.1016/j.copbio.2014.11.007.

Bio Suisse. CRISPR/Cas und die Biobranche. Accessed December 20, 2017 from www.bio-suisse.ch/media/Ueberuns/Medien/unsere_meinung_zu_crispr_cas.pdf.

Bioland. Gentechnikverfahren CRISPR/Cas ist absolutes No-Go für Biolandwirtschaft: Leben ist nicht programmierbar. Accessed December 20, 2017 from www.bioland.de/presse/presse-detail/article/bioland-leben-ist-nicht-programmierbar-gentechnikverfahren-crisprcas-ist-absolutes-no-go-fuer-biolandwirtschaft.html.

Blok, V. and Gremmen, B. 2016. Ecological innovation: Biomimicry as a new way of thinking and acting ecologically. *Journal of Agricultural and Environmental Ethics* 29, no. 2: 203–17.

Brooks, C., Nekrasov, V., Lippman, Z. B. and Van Eck, J. 2014. Efficient gene editing in tomato in the first generation using the clustered regularly interspaced short palindromic repeats/CRSIPR-associated9 system. *Plant Physiology* 166, no. 3: 1292–7.

Bruce, D. M. 2002. A social contract for biotechnology: Shared visions for risky technologies? *Journal of Agricultural and Environmental Ethics* 15, no. 3: 279–89.

Carroll, D., Van Eenennaam, A. L., Taylor, J. F., Seger, J. and Voytas, D. F. 2016. Regulate genome-edited products, not genome editing itself. *Nature Biotechnology* 34, no. 5: 477–9.

Dabrock, P. 2009. Playing god? Synthetic biology as a theological and ethical challenge. *Systems and Synthetic Biology* 3, no. 1–4: 47–54.

European Parliament. 2001. EU Directive 2001/18/EC of the European Parliament and of the Council of 12 March 2001 on the deliberate release into the environment of genetically modified organisms and repealing Council Directive 90/220/EEC. *Official Journal of the European Communities L* 106: 1–39. Accessed December 20, 2017 from http://eur-lex.europa.eu/legal-content/EN/TXT/?uri=CELEX:32001L0018

Huang, S., Weigel, D., Beachy, R. N. and Li, J. 2016. A proposed regulatory framework for genome-edited crops. *Nature Genetics* 48, no. 2: 109–11.

IFOAM. 2014. The IFOAM Norms for organic production and processing. Version 2014, IFOAM, Germany, Accessed December 2, 2017 from www.ifoam.bio/sites/default/files/ifoam_norms_version_july_2014.pdf.

IFOAM EU. 2015. *New plant breeding techniques – Position paper*. Brussels: IFOAM EU. Accessed December 20,2017 from www.ifoam-eu.org/sites/default/files/ifoameu_policy_npbts_position_final_20151210.pdf.

Jacobs, T. B., LaFayette, P. R., Schmitz, R. J. and Parrott, W. A. 2015. Targeted genome modifications in soybean with CRISPR/Cas9. *BMC Biotechnology* 15: 16. https://doi.org/10.1186/s12896-015-0131-2.

Karafyllis, N. C. 2003. Das Wesen der Biofakte. Pp. 11–27 in *Biofakte: Versuch über den Menschen zwischen Artefakt und Lebewesen*, edited by N.C. Karafyllis. Paderborn: Mentis.

Kim, J. and Kim, J. S. 2016. Bypassing GMO regulations with CRISPR gene editing. *Nature Biotechnology* 34, no. 10: 1014–5.

Koornneef, M. and Stam, P. 2001. Changing paradigms in plant breeding. *Plant Physiology* 125, no. 1: 156–9.

KWS. 2017. New plant breeding methods. Accessed December 3, 2017 from www.kws.com/aw/Methods/New-Plant-Breeding-Methods/~hjmg/.

Lammerts van Bueren, E. T., Struik, P. C., Tiemens-Hulscher, M. and Jacobsen, E. 2003. Concepts of intrinsic value and integrity of plants in organic plant breeding and propagation. *Crop Science* 43, no. 6: 1922–9.

Lee, K. 2003. *Philosophy and revolutions in genetics: Deep science and deep technology, renewing philosophy*. Basingstoke, UK: Palgrave Macmillan.

Nuijten, R. J. M., Bosse, M., Crooijmans, R. P. M. A., Madsen, O., Schaftenaar, W., Ryder, O. A., Groenen, M. A. M. and Megens, H.-J. 2016. The use of genomics in conservation management of the endangered visayan warty pig (*Sus Cebifrons*). *International Journal of Genomics* 2016: 9.

Nuijten, E., Messmer, M. M. and Lammerts van Bueren, E. T. 2017. Concepts and strategies of organic plant breeding in light of novel breeding techniques. *Sustainability* 9, no.1:18.

Palmgren, M. G., Edenbrandt, A. K., Vedel, S. E., Andersen, M. M., Landes, X., Osterberg, J. T., Falhof, J. et al. 2015. Are we ready for back-to-nature crop breeding? *Trends in Plant Science* 20, no. 3: 155–64.

Schouten, H. J., F. A. Krens and E. Jacobsen. 2006. Cisgenic plants are similar to traditionally bred plants: International regulations for genetically modified organisms should be altered to exempt cisgenesis. *EMBO Reports* 7, no. 8: 750–3.

Siipi, H. 2008. Dimensions of naturalness. *Ethics and the Environment* 13, no. 1: 71–103.

Steinbrecher, R. 2015. Genetic engineering in plants and the "New Breeding Techniques" (NBTs) – Inherent risks and the need to regulate. Accessed December 3, 2017 from www.econexus.info/sites/econexus/files/NBT Briefing – EcoNexus December 2015.pdf.

Svitashev, S., Young, J. K., Schwartz, C., Gao, H., Falco, S. C. and Cigan, A. M. 2015. Targeted mutagenesis, precise gene editing, and site-specific gene insertion in maize using Cas9 and guide RNA. *Plant Physiology* 169, no. 2: 931–45.

Maurin, J. 2016. Ökoforscher über neue Gentech-Methode: "CRISPR hat großes Potenzial". *Die Tageszeitung*, April 6. Accessed December 20, 2017 from www.taz.de/Archiv-Suche/!5290509&s=CRISPR/.

Then, C. 2016. *Synthetic gene technologies applied in plants and animals used for food production: Overview on patent applications on new techniques for genetic engineering and risks associated with these methods*. Munich: Grassroots Foundation.

Van Haperen, P. F., Gremmen, B. and Jacobs, J. 2012. Reconstruction of the ethical debate on naturalness in discussions about plant-biotechnology. *Journal of Agricultural and Environmental Ethics* 25, no. 6: 797–812.

Weigel, D. 2017. Ethische Fragen spielen keine Rolle – Einsatz Chancen und Risiken von CRISPR/Cas bei Pflanzen. *Forschung und Lehre* 1, no. 2017: 22–23.

Woo, J. W., Kim, J., Kwon, S. I., Corvalan, C., Cho, S. W., Kim, H., Kim, S. G., Kim, S. T., Choe, S. and Kim, J. S. 2015. DNA-free genome editing in plants with preassembled CRISPR-Cas9 ribonucleoproteins. *Nature Biotechnology* 33, no. 11: 1162–4.

Zwart, H. 2009. Biotechnology and naturalness in the genomics era: Plotting a timetable for the biotechnology debate. *Journal of Agricultural and Environmental Ethics* 22, no. 6: 505–29.

18 "Digital" plants and the rise of responsible precision agriculture

Bart Gremmen and Vincent Blok

Introduction

Ever since agriculture began, plants have been at the center of an innovation trajectory aimed at improving their use for human consumption. Starting with a huge diversity of plant species, and a vast inclusion of genetic traits in the production process, some plants were slowly domesticated into crop plants through a breeding endeavour based on trial and error. The end result of thousands of years of selection in the developed world is an enormous extension of cultivated fields, with only a few kinds of uniform crop plants that are produced in monocultures.

Over the last few hundred years, there has been a plethora of innovations in crop-based farming techniques. Although agricultural mechanization, hybrid cultivars, the Green Revolution, and modern biotechnology have been implemented on a global scale and have become very important in feeding the world population, mainstream farmers face much criticism, including normative criticism, from organic farmers and various kinds of non-governmental organizations. The main points of criticism are: pesticide and fertilizer use, soil degradation, and use of genetically modified plants.

Because plant ethics is largely under-theorized, the arguments moved by critics to conventional farming are mainly derived from animal ethics and are modeled on arguments on individual animals. For instance, critics argue that individual plants are not just instruments for profit-driven enterprises, but also have an intrinsic value that commands respect (Rollin 1992). Mainstream farmers are thus urged to reorganize their practices in a responsible way and no longer consider plants as mere instruments or objects. This focus on individual plants leads to application problems, because farmers are dealing with crops, not individual plants.

By embracing precision agriculture (PA), the latest technological innovation, mainstream crop production seems to be pushing further in the direction of large-scale cultivation, now through large-scale digitalized farming. At the same time, PA technologies can also be said to create, by integrating large numbers of diverse data, so many individual "digital" plants, each distinct and identifiable. This raises ethical issues regarding the changed farmer–plant relationship.

Our goal in this chapter is to investigate, from the perspective of plant ethics, the potential to develop a responsible PA, and the limitations to such an enterprise. We articulate an applied plant ethics framework centered on the notion of plant "flourishing" (Kallhoff 2014). Flourishing can be defined as the good life of plants, and has four dimensions: (1) a plant is viable throughout its life (the relative strength of a plant in a particular situation); (2) a plant is capable of accomplishing its typical life cycle; (3) a plant succeeds in expressing the typical characteristics of the specific organism with the specific life-form that it is, generally fitting its species description; and (4) a plant succeeds in performing its role in a specific eco-system.

We will proceed as follows: briefly sketch a history of PA (Section Two), describe a flourishing-centered ethical framework for plants (Section Three), and consider it in connection to the following key technologies that have influenced PA's development: sensing technologies, data processing and utilization, and control technologies (Section Four). Finally, we discuss the possibility of developing a responsible precision farming of crop plants, based on a plant ethics of flourishing.

Precision agriculture

PA refers to a cluster of technological advancements that have already been incorporated into crop production. PA is a form of digitalization comprising a set of technological tools combined with data-based models that rely upon automated monitoring of biological and physical processes (Wathes et al. 2008). It is made possible by new technologies, such as geographic positioning systems (GPS), sensors, geographic information systems (GIS), and advanced software and precision application equipment. It is the goal of PA to modify inputs (e.g. fertilizer, irrigation, dairy effluent, seed rate) spatially and temporally at the subfield scale in view of cost efficiency, productivity, and environmental gains. Globally, the affordability and accessibility of these technologies helped PA emerge as a research discipline in the 1980s (Zhang, Wang, and Wang 2002). Strong emphasis has always been put on improving nutrient use efficiency, by matching inputs to site-specific field conditions.

PA treats crop production as a set of interlinked processes, which act together in a complex network. For example, plant growth is a process suitable for a PA approach, but also reducing the amount of seeds needed, and the prevention of some endemic diseases are similarly suitable. PA may also be concerned with the larger physical environment of a field, such as its thermal microenvironment and emissions of gaseous pollutants (Bramley 2009).

PA is an all-compassing term for the use of a set of technologies that promote improved management of agricultural production, following a recognition that the productivity of agricultural land can vary considerably, even over very short distances. PA can be regarded as a new way of controlling the growth of a crop, and a means to increase the chance that the right crop management strategies are implemented at the right time and in the right place (Bramley 2009). PA can be applied to individual plants, a crop, or a whole field (Wathes et al. 2008). The level of application also correlates with the degree of control by the farmer, varying from a basic level of data collection and registration to more comprehensive levels that pertain to process management, product management, and business intelligence.

Progress in PA has been facilitated by significant improvements in computer processing power and the availability of different sensor technologies that are mainly "borrowed" from other industries (Bramley 2009). Common features of sensors that are currently used in PA are continuous, simultaneous measurement of a large number of units with analysis of temporal trends as well as point statistics (Wathes et al. 2008). The concept of PA as a fully integrated system still needs to be further developed: the most pressing needs are data-based models with meaningful parameters, and control systems that can manage two or more interacting biological and/or physical processes.

An ethical framework for digital plants

Before proceeding to a normative evaluation of some emerging technologies falling under the general rubric of PA, we need an ethical framework to provide us with assessment criteria. In applied ethics, the "ethical matrix" is often used as a process tool to develop such a framework (Mepham et al. 2006). The matrix consists of a horizontal row of three *prima facie* principles: respect for welfare, autonomy, and fairness. The vertical row of the matrix consists of a list of relevant stakeholders that is the set of selected interest groups for which the three principles have to be applied.

In Mepham's ethical matrix, plants are considered as biota. When it comes to biota, Mepham translates his three principles as follows: welfare becomes conservation, autonomy is understood as biodiversity, and fairness is translated as sustainability (Mepham et al. 2006). This approach, however, is inadequate because it is too general, for two reasons: first, we are looking for an ethical framework for plants. Since "biota" is the totality of all organisms of a region or a time of day, all plants are biota, but not all biota are plants. It does not do justice to the ethical differences between living beings to lump animals, fungi, bacteria, viruses and plants all under the catch-all term "biota". Second, the concepts of "conservation", "biodiversity", and "sustainability" are still very broad and do not provide sufficient guidelines for ethical considerations in agricultural practices.

A. Kallhoff's (2014) flourishing-centered approach to plant ethics improves upon Mepham's matrix by translating the concept of plants welfare into that of flourishing (as opposed to conservation, as Mepham himself does). Flourishing as the pivotal concept is a pre-moral and yet evaluative notion about the life of plants (Kallhoff 2014). It describes the ethical "good" of a plant and enables us to distinguish between effects that are beneficial and effects that are harmful to plants.

According to Kallhoff (2014), a plant flourishes when all of the following conditions occur:

1. "It [the plant] is viable throughout its life, so that it is capable of reacting to external stress without endangering the overall performance which sustains its life" (Kallhoff 2014, 687). Viability is a concept that explains the relative strength of a plant in a particular situation. Unlike "health", viability implies that a plant is not only free from an "illness" but is also in the position to perform its vital functions successfully.
2. "It is capable of accomplishing its typical life-cycle (juvenile phases and adult phases which end with proliferation)" (Kallhoff 2014, 687). This condition indicates that flourishing is a concept relating to the goodness of the whole life of an individual plant. Plants are organized such that the achievement of a complete life cycle has priority over, for example repairing damaged roots or leaves. Proliferation is an important element in the realization of flourishing.
3. "It succeeds in expressing the typical characteristics of a plant which has a specific life-form" (Kallhoff 2014, 687). This condition highlights the importance of a plant being a good token of its kind.

Kallhoff considers these three conditions to suffice for plant flourishing, but we suggest the importance of a fourth condition:

4. It succeeds in performing its role in a specific ecosystem. Most plants are literally rooted in their surroundings. Individual plants cannot survive without a specific interdependency of fungi, birds, fertile soil, insects, other plants, etc. (cf. Blok 2016).

In the next section, we will use the four conditions of plant flourishing as our ethical framework for evaluating the moral quality of three key technologies in PA: sensing technologies, data processing, and control technologies.

Key technologies in PA

Sensing technologies

PA started in the 1980s with the development of a high-precision positioning system (Liaghat and Balasundram 2010). Currently, many satellites enable navigation and positioning capability within the agricultural field anywhere on Earth, anytime, and under any conditions. Geographic coordinates (latitude and longitude) are used to record the position of the field, and also to navigate agricultural vehicles within the field, with a 2 centimeters accuracy (Mahlein 2016). Satellites produce maps holding information about soil type, nutrients levels, and so on, and assign that information to the particular field location. Drones and sensors collect data to evaluate soil and crop health.

Although farmers have always collected huge amounts of data by traditional means, PA entails a switch from analogous to digital data and expands the quantity, quality, and range of the data. Sensors can be applied to any process that is part of farming (Mahlein 2016): the level of application depends mainly on the economics of the crop. For instance, low-cost cameras in combination with image analysis can be used to quantify the size and shape of the plants in the field, even when it comes to large crops. Sensors can also be directly placed on the plant, thereby placing the individual plant at the center of PA (Wathes et al. 2008). Farmers will even be able to take care of individual plants.

How do we evaluate sensor technologies that are used in PA while keeping the flourishing of the plants and crops as yardstick? First, compared to the classical, physical and visual inspection of crops done by the farmer in a field, sensor technologies disturb plants less, so plants are still in a position to perform their vital functions successfully. In the past, farmers would inspect their crops by literally walking through the plants, while PA uses the electromagnetic spectrum to gather data, and thus does not disturb the ecosystem of the plants grown in mono-cultures. This enables farmers to recognize the presence of external stressors without endangering the overall performance of plants, and this sustains rather than hinder their life, satisfying the first condition of our flourishing-centered ethical matrix.

Second, in contrast with former management strategies, sensor technologies provide important data that will enable farmers to accurately administer just the right amount of manure, pesticides, and water to assure that the plants are in a position to express the typical characteristics of plants that have a specific life-form, thus satisfying the third condition of our ethical matrix. In addition, the data of sensor technologies could help improve the typical life cycle of the plants involved, thus meeting the second condition in our matrix. However, PA applied to monocultures does not strengthen the ecosystemic interdependence of plants, which is the fourth condition of our ethical framework. PA compensates for this lost ecosystemic interdependence of plants by improving their management and control.

Data processing

PA has the advantage of collecting large amounts of information, and the quality of the information is very high. However, the raw data are about specific aspects of plant life only. Previously, the gaze of the farmer on the field and extensive experience immediately integrated various aspects of plant life into a single, useful picture. In PA this task is taken over by softwares processing the data according to complex models (Zhang, Wang, and Wang 2002), so that scientific information is needed to analyze and interpret the data provided by sensor technologies. But assuming that, PA that is able to integrate different aspects of plant life can surely strengthen the overall performance of a plant which sustains its life, satisfying the first condition of our flourishing-centered ethical matrix.

One of the main requirements to integrate all components in PA is integrated electronic communications between all the components in a farming system. Interoperability is a key success factor in digital farming, because connecting agricultural machines is only possible when farm machines can seamlessly communicate with each other and to related systems (such as farm management systems). This is often described as the "internet of things"; it links all kinds of machines to the farm office and also to stakeholders in the context of the farm, such as dealers of supplies (Bramley 2009). The "internet of things" assures that all the available resources are mobilized in time to enable the expression of all the typical characteristics of plants that have a specific life-form, thus satisfying the third condition of our ethical matrix.

When integrated, the data models used in PA to process the data could be interpreted as virtual life cycles of plants in the fields. This could help improve the typical physical life cycle of the plants involved, thus meeting the second condition in our matrix. The information about the actual physical and biological processes in the plants is supplemented by information about the ecosystem of the plants, the actual and historical soil, weather, and environmental conditions, but PA does not use this information to strengthen the interdependence of plants in the ecosystem, and thus it does not meet the fourth condition in our matrix.

Control technologies

Control technologies in traditional agriculture depend on the farmer's ability to steer machinery on the land. They tend to hamper the flourishing conditions of crop plants in a number of ways. For example, land can be spoiled by the inefficient turning of machinery; uneven rows may be created; and typically too many or too few resources (such as seeds, manure, or pesticides) may be expended on a certain location. New control technologies offer solutions to these problems and improve the life cycle of plants and agricultural ecosystems (Lu et al. 2016), thus meeting the second condition. Again, though, it does not meet the fourth: the solutions afforded by PA to the traditional control problems mentioned above limit the farmer's inefficient behaviour, but they do not strengthen the ecosystemic interdependence of plants.

The "internet of things" is key to not only integrated electronic communications between all the components in farming systems, but also to the use of the information by control technologies (Darnhofer, Gibbon, and Dedieu 2012). Automated steering systems enable the takeover of specific driving tasks. These technologies not only aim to reduce human error but also strive to achieve effective site management (Lu et al. 2016). This strengthens the overall performance of plants and the expression of all the typical characteristics of plants that have a specific life-form, thus satisfying the first and third conditions of our ethical matrix.

The ethics of digital plants

From the perspective of plant ethics, what are the possibilities and limitations related to developing a responsible PA? We started with the ethical matrix of

Mepham and adapted it for the purposes of plant ethics so that the concept of flourishing, as developed in Kallhoff (2014), was central to the evaluation. Because PA is able to redirect focus from crops to individual plants, in order to evaluate PA technologies that effectively "digitalize" plants we can use a flourishing-centered ethical matrix that applies to individual plants. The flourishing of plants is usually reflected in PA at a particular level of crop health, environmental protection and/or efficient use of natural resources.

When does PA qualify as responsible? We propose that PA may be considered responsible when all the conditions of the flourishing-centered ethical matrix are met. We want to prevent that PA enables farmers to endanger the flourishing of plants by only focusing on aspects of their data processing and steering technologies that ensure high productivity and high profits. A responsible PA also implies changes in what is considered a good farmer and what is considered an appropriate relationship between the farmer and the plant. PA promises to ensure the application of best-practice farming, and is thus likely to reduce variability in farming, particularly variability in inputs and output (Zhang, Wang, and Wang 2002). This raises the question whether PA is an example of a technological intervention that can be associated with the unsettling of traditional, and possibly valuable, farmer–plant relationships and ethical identities.

Is PA really changing the farmer–plant relationship, though? To reduce costs, PA substitutes labor with various electronic devices that become the hands, the ears, and the eyes of the farmer. This means that farming activities such as sowing, plowing, and monitoring of plants become almost fully automated, which simplifies the work of farmers. Thanks to PA, farmers are likely to devote less time and attention to caring for the health and welfare of their "digital" plants. Yet, this is not harmful to the plants, because sensors will monitor plant health and welfare every hour of the day, and more precise automated administration of inputs will sustain them. The different control technologies will also ensure that "digital" plants are capable of accomplishing their typical life cycle and succeed in expressing their typical characteristics. The seeds will be placed at the right spot in the soil, will get the right amount of water and nourishment, and will be protected from pests. Humans will manage their "digital" plants from control rooms rather than on the ground.

Responsible PA in mono-culture farming changes the farmer–plant relationship but does not disrupt it. In contrast to the flourishing conditions of plants in the wild, by meeting the flourishing conditions of plants in agriculture, and thus meeting the first three conditions of our flourishing matrix, PA is primarily used to ensure economic gain by efficient production.

This could stimulate crop production, but still leaves the fourth condition to be met. However, the interdependency of plants in a mono culture farming system is very difficult to achieve. This means that a responsible PA, in contrast to plants in the wild, will restrict the flourishing conditions of 'digital' plants in agriculture by controlling their proliferation and diversity and, will enhance the flourishing conditions of 'digital' plants by reducing many uncertain factors that hamper wild plants.

References

Blok, V. 2016. Biomimicry and the materiality of ecological technology and innovation in advance: Toward a natural model of nature. *Environmental Philosophy* 13, no. 2: 195–214.

Bramley, R. G. V. 2009. Lessons from nearly 20 years of precision agriculture research, development, and adoption as a guide to its appropriate application. *Crop and Pasture Science* 60, no. 3: 197–217.

Darnhofer, I., Gibbon, D. and Dedieu, B. 2012. Farming systems research: An approach to inquiry. Pp. 3–31 in *Farming systems research into the 21st century: The new dynamic*, edited by I. Darnhofer, D. Gibbon and B. Dedieu. Dordrecht: Springer Netherlands.

Kallhoff, A. 2014. Plants in ethics: Why flourishing deserves moral respect. *Environmental Values* 23, no. 6: 685–700.

Liaghat, S. and S. K. Balasundram. 2010. A review: The role of remote sensing in precision agriculture. *American Journal of Agricultural and Biological Sciences* 5, no. 1: 50–55.

Lu, C., Meng, Z., Wang, X., Wu, G., Gao, N., and Dong, J. 2016. Application feasibility analysis of precision agriculture in equipment for controlled traffic farming system: A review. Pp. 348–55 in *Computer and computing technologies in agriculture IX*, edited by Li, D. and Li, Z. Cham: Springer International Publishing.

Mahlein, A.-K. 2016. Plant disease detection by imaging sensors – Parallels and specific demands for precision agriculture and plant phenotyping. *Plant Disease* 100, no. 2: 241–51.

Mepham, B. 2000. A framework for the ethical analysis of novel foods: The ethical matrix. *Journal of Agricultural and Environmental Ethics* 12, no. 2: 165–76.

Mepham, B., Kaiser, M., Thorstensen, E., Tomkins, S. and Millar, K. 2006. *Ethical matrix manual.* The Hague: LEI.

Palmer, C. 2008. Environmental ethics and agricultural intensification. Pp. 131–48 in *The ethics of intensification*, edited by P. Thompson. The Netherlands: Springer.

Rollin, B. E. 1992. *Animal rights and human morality*. Buffalo, NY: Prometheus Books.

Wathes, C. M., Kristensen, H. H., Aerts, J. M. and Berckmans, D. 2008. Is precision livestock farming an engineer's daydream or nightmare, an animal's friend or foe, and a farmer's panacea or pitfall? *Computers and Electronics in Agriculture* 64, no. 1: 2–10.

Zhang, N., Wang, M. and Wang, N. 2002. Precision agriculture—A worldwide overview. *Computers and Electronics in Agriculture* 36, no. 2: 113–32.

19 On robots and plants

The case of the plantoid, a robotic artifact inspired by plants

Barbara Mazzolai and Pericle Salvini

Introduction

Since ancient times, human beings have made use of techniques and technologies for mastering and exploiting nature, including plants (e.g. crop rotation, grafting, greenhouses, threshing machines). Robotics, which is the latest and most advanced form of technology developed by humans, is increasingly being applied to agriculture, improving results in terms of efficiency of production as well as sustainability (King 2017). In particular, unmanned aerial vehicles (UAVs), better known as drones, are now being designed and used for a variety of important agricultural purposes: soil and field analysis, planting, crop-spraying and monitoring, irrigation, and more (Mazur 2016).

The subject of this chapter, however, is not new technologies that can be applied to plants but rather technologies that are modeled on plants. The Plantoid, a robot inspired by plants, uses plants as models to discover scientific facts and principles that can lead to the development of new technologies, while at the same time increasing our understanding of the world of plants.

Plants have inhabited Earth longer than most other living beings. Over billions of years of evolution, plants had to develop smart and efficient adaptation strategies as well as ways to reduce energy consumption during motion. In addition to these adaptation and energy-saving techniques, studying plants can help us develop new smart materials, autonomous systems to environmental exploration and monitoring, and new sensor techniques. More generally, plants can provide innovative insights for technologies that may be of great importance in the future, particularly if we investigate their responses to changing environmental conditions (such as air humidity) or learn to mimic some of their characteristic processes (such as the ability to grow for the entire life).

There are many examples of technological solutions inspired by plants. In some cases, these solutions have already translated into commercial products. A famous plant-inspired solution is Velcro®, which is a hook-and-loop fastener used in many everyday applications. It was developed in 1948 by a Swiss engineer, George de Mestral, who observed that the hooks in plant burrs (*Arctium lappa*) remained attached to the fur of his dog. In 1955, he patented Velcro®, thereby spurring the creation of a variety of other products based on this simple yet brilliant innovation.

Other plant-inspired technological solutions are related to innovative materials that imitate specific plant functionalities. The super-hydrophobic and self-cleaning properties of leaves of water-repellent plants, such as *Nelumbo nucifera* (lotus) and *Colocasia esculenta*, are at the basis of the so-called "lotus effect", a result of hierarchical roughness (i.e. microstructures formed by papillose epidermal cells covered with epicuticular wax tubules) and the presence of a hydrophobic coating (Bhushan 2009). Many artificial solutions (e.g. StoColor Lotusan® and new micro- and nano-patterned polymeric- or graphene-based materials) have been developed based on the study of these plants' features (Zang et al. 2013). New materials have also been generated by the study of *Nepenthes* pitcher plants and their repellent surfaces: for example Wong et al. (2011) recently developed "slippery liquid-infused porous surfaces" (SLIPS) that consist of a film of lubricating liquid, held in place by a micro- or nano-porous substrate.

Much like these innovations, the Plantoid robot is bio-inspired. It is a robot designed by taking inspiration from nature. The basic idea of biorobotics is to study nature to generate new knowledge and to develop new technologies (Young et al. 2016).

In this chapter, we shall describe the Plantoid and introduce the reader to the field of plant-inspired robotics. The ethical dimensions of plant-inspired robotics pertain both to robot and to plant ethics. Plant ethics is concerned, among other things, with the study of "the relations between plants and persons".[1] The Plantoid can contribute to this enterprise, raise awareness about the complexity of plant life, and educate people to more sophisticated uses of such fundamental force in our ecosystems. Later in the chapter we shall highlight how that intersects with issues in robot ethics, thus creating a somewhat unexpected new field of interdisciplinary research.

The chapter is organized as follows: in Section 1, we describe the design and development of the Plantoid and its capabilities; in Section 2, we discuss the Plantoid from an ethical standpoint.

The Plantoid: A robot that grows, perceives and communicates

As its name suggests, the Plantoid is a robot in the form of a plant. It is a robot because it has a central processing unit and is made of sensors and actuators that make it capable of growing, perceiving and communicating autonomously. But most relevantly it is a robot because it is a built machine whose purpose is to serve human needs, and which can be programmed to do so (e.g. to explore terrestrial as well as extra-terrestrial soils). Etymologically, the word "robot" comes from the Slavonic "robota", which means "drudgery" and "forced labor".[2]

The Plantoid also responds to deeply-rooted demiurgic ambitions that humans have displayed through the centuries and that are specifically devoted to the creation or replication of life-forms (think of the Golem, an animated anthropomorphic being that, in Jewish folklore, is magically created entirely from inanimate matter, specifically clay or mud – just like humans themselves were created by God in the Old Testament). The Plantoid is an attempt to imitate a real plant in its functions. It is, perhaps surprisingly, the first attempt of this kind: so far, the imitation of nature by means of robotic counterparts has strangely been limited to human and animal species (Pfeifer, Lungarella and Iida 2007; Ijspeert 2014).

More specifically, the Plantoid is an artificial replica of plant roots – and the first robot worldwide that imitates the principles that enable plant roots to efficiently adapt to underground environments. Roots are the organs delegated

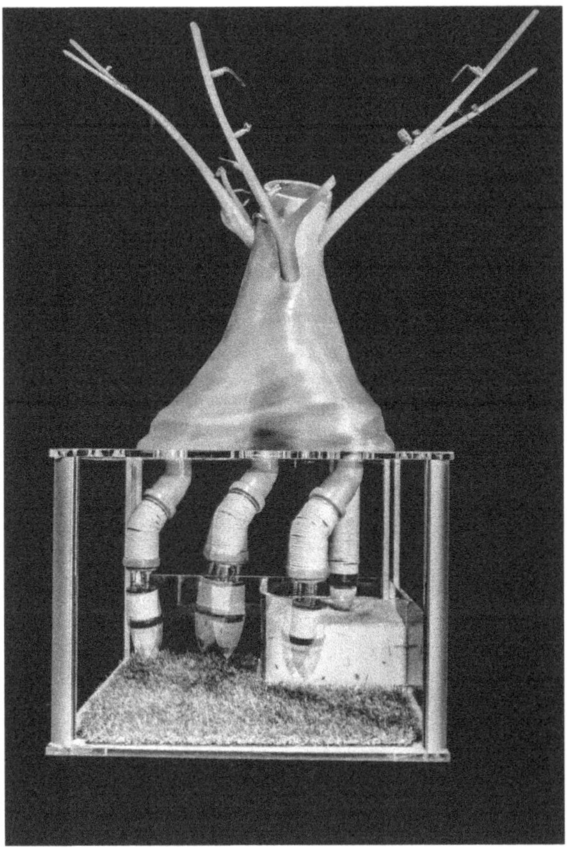

Figure 19.1 Picture of the plantoid prototype that includes a trunk, used to collect data from the roots and to exchange data with an external PC, several branches with artificial leaves and a total of five roots. The root apex is the same for all the roots, while the actuation system varies: there are one root for growing and four for bending (Sadeghi et al. 2016)

Photo credit: Marco Moscato.

to the foraging and the anchoring of plants. While performing these tasks, roots – whose apices are highly sensorized and explore soil in capillary fashion – need to adapt to the environment, avoid obstacles, penetrate soil, mine minerals and water to fulfill their primary functions (Hart 1990). The responses of the roots to environmental interactions and stimuli are called *tropisms*. Roots show a considerable plasticity in their morphology, in response to the variability of the chemical–physical conditions or the environment in which they move.

The Plantoid is composed of a trunk, five functional roots that originate from the trunk with sensorized apexes (or tips), and an aerial part with branches, including polymeric artificial leaves (based on controllable hygromorphic plant-inspired material moving in response to humidity) (Taccola et al. 2015). The Plantoid is equipped with several sensors, including soft and flexible touch and humidity sensors for pH, potassium, nitrate and phosphate. Analogous to its natural counterpart, each robotic tip integrates all these and other sensors, and this makes for an intelligent robotic root capable of monitoring the environment that it is exploring (Sadeghi et al. 2016) (Figure 19.1).

A main achievement was to identify how to mimic the growing and bending process of plant roots. Roots grow to move towards or far from an environmental stimuli. This phenomenon is mediated by two continuous processes in the apex: cell division and cell elongation, in the meristematic and elongation zones, respectively (Verbelen 2006). The symmetric growth of cells at the root sides results in straight penetration of the root into the soil, while asymmetric cell growth (or division) is responsible for bending and steering behaviours (differential growth/elongation).

The Plantoid can grow at apex level by addition of new artificial material in a process that resembles that of 3D printing. Each robotic root includes a tubular body, a growing head and a sensorized tip that commands the robot behaviors. The growing head is a customized 3D printer-like system that builds the tubular body of the root in the format of circular layers, by fusing and depositing a thermoplastic material (i.e. polylactic acid [PLA] filament) at the tip level. In this way, movement is obtained by growing. A differential deposition of the material can create an asymmetry that results in curvature of the built structure, providing the possibility of root bending to follow or escape from a stimulus, or to reach a given point in space. Taking advantage of these characteristics, the robotic roots are able to move inside a medium by growing their body (Sadeghi et al. 2017) (Figure 19.2).

In nature, each root apex can be considered as a "command center" at which level the plant perceives the environment. It then directs the root growth accordingly, and also implements strategies of competition, defense and relation to other plants or organisms. This determines an emergent behavior. Analogously, in Plantoid, each artificial root has its own controller to implement behavior. Such controller consists of two sub-systems: a root apex and a root driver. The root apex is dedicated to data acquisition from the sensors and to the implementation of the corresponding behavior (tropism), based on a high-level algorithm that defines the growth direction. The root driver implements the low-level

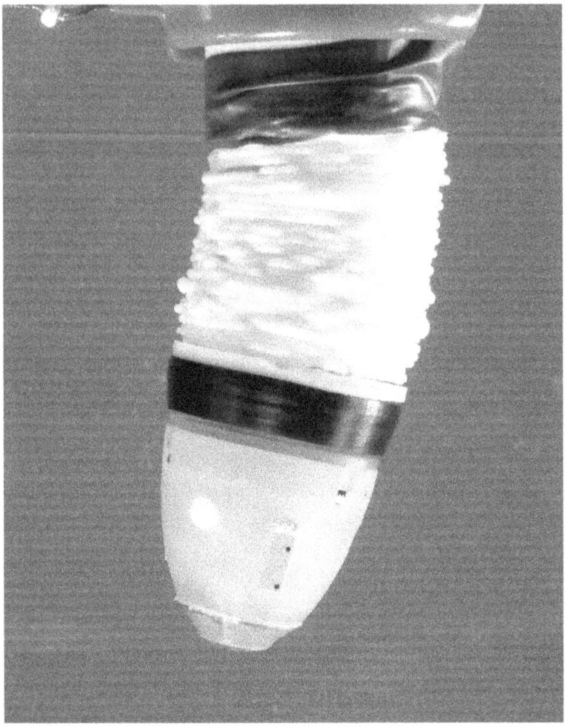

Figure 19.2 An artificial growing robotic root. The picture shows the growing mechanism with the sensorized tip. The tubular body is made of PLA, poly-lactic acid.

Photo credit: Ali Sadeghi, Alessio Mondini.

control by driving the actuators for the growing and bending roots (Sadeghi et al. 2016).

These and other abilities have a plethora of applications. The mark of these applications is that they are typically muscle-free and thus quite different from what may be inspired by animal studies and animal-based modelling. Plants represent an excellent paradigm for robotics in terms of energy efficiency, low power consumption and low friction and low speed movements; this makes them particularly suitable for operating in narrow and unstructured environments (Mazzolai 2017).

Applications of Plantoid include primarily environmental monitoring and soil exploration for nutritional substances, water, as well as potential pollutants, whether on earth or other planets. Other applications are possible in the medical and surgical field, for example, where the strategies of plant roots and their robotic replicas can inspire the development of new flexible endoscopes, able to steer by growing in delicate human organs (Mazzolai, Beccai and Mattoli 2014).

The ethics of robots

The interest towards robot ethics or roboethics[3] consolidated around the years 2003–2005. A milestone event was the First International Symposium on Roboethics in 2004,[4] in which a multidisciplinary group of scholars – philosophers, jurists, sociologists, anthropologist, ethicists and robot engineers – discussed topics such as the rights of robots, lethal applications of robots, moral dilemmas such as whether robots can be better than humans, and legal liability. Nowadays, robot ethics has become a consolidated research area within the robotics community,[5] attracting scholars from all disciplines – as demonstrated by the frequent sessions or workshops on roboethics hosted in prime robotics conferences,[6] and the Technical Committee in the Robotics and Automation Society of IEEE.[7]

Since its origins, roboethics has been an umbrella term used for including interdisciplinary research on the broad implications of robotics, covering not just ethical but also legal, societal, economic and political issues. Roboethics is not meant to hinder technological and scientific progress: on the contrary, its purpose is to encourage and guide progress towards sustainability, both in relation to human beings and the environment, and looking at both the present and the future.

Although every kind of technology may have inherent value-laden characteristics, which are independent from how well or badly humans use it, roboethics is mainly concerned with the human agent acting as designer or user. According to this conception, and in line with RRI,[8] ethics should apply to both the innovation process (deontological ethics) – it should be concerned with the code of conduct of researchers or professional engineers – and with the use of a robotic product (consequentialist or virtue ethics), namely with the morality of robot use. No code of ethics yet exists for professional robot engineers, so the ethics of robotics has been mostly consequentialist, focusing on the evaluation of the ethical implications of certain robot applications.

Overall, it is possible to maintain that roboethics has two main goals. One is to promote and favor applications that can improve the quality of life of people and the conditions of the natural environment – as, for instance, with biomedical and precision farming applications. The other is to minimize the risks and dangers to people's health and well-being that are connected with robotics research and applications – for instance, when it comes to robots developed for military purposes. The evaluation of potential risks and dangers should address both physical as well as psychological hazards.

Research on risk assessment and management when it comes to robots has focused mainly on trying to identify and solve so-called "hard" risks of human–robot interactions – that is, dangers emerging from physical impacts between robots and human beings, such as injures deriving from clashes between a robotic arm and a person (Bicchi, Peshkin and Colgate 2008). However, the recent trend towards developing sociable or companion robots, in which machines designed with an anthropomorphic or zoomorphic appearance are endowed with social and even emotional capabilities, may trigger new risks and vulnerabilities, which may be called "soft".

Contrary to hard risks, soft ones are caused by the normal functioning of a robot, and they consist of hazards related to psychological impacts on the user. These impacts can affect a person's cognitive, social and affective health. Examples of soft risks are those related to robotic or digital devices that replicate human or animal behaviors (e.g. robotic pets or chatbots) and may foster "subconscious engagement" (a compulsion to care for artifacts that exhibit social capabilities – cf. Turkle 2007; 2012); and those related to the effects of mediation, which can affect human cognitive abilities, as reported by studies on aircrafts' pilots suggesting that "automation complacency and bias" can determine de-skilling effects, and compromise attention and awareness during a task execution (Noyes, Cook and Masakowshi 2007; Parasuraman, Molloy and Singh 1993). But these concerns do not seem to apply to the Plantoid in any obvious way. Of course, as for any technology, its impact cannot easily be predicted before the technology has fully developed and become socially pervasive (Collingridge 1980). Still, at this point it seems that the Plantoid can only improve our understanding of plant life, increase our respect for it, and deepen the human-vegetal relation.

Conclusions

In this chapter, we presented an innovative robot inspired by a plant, capable of sensing and acting. We discussed some potential, ethically troubling implications of its applications. We found that classic worries in robot ethics do not seem to apply to plant-inspired robots. But the ethics of bio-inspired robots is in its infancy, so that even fathoming potential ethical, political and economic issues is difficult. Our alertness to them should remain high, however, and must extend to both the research process and the realms of expected applications.

For now, however, we would like to highlight the educational value of the Plantoid robot. For centuries, plants have been considered passive beings, inferior to other living entities not just because they are incapable of moving but also because they have neither intelligence nor feelings. This negative perception of plants is also affected by technical obstacles to the observation of their movements (especially of the roots growing in the soil) and to the study of their sensing and decisional capabilities. This has also caused a delay in the adoption of plants as models for developing robots for which movement, perception and control are essential.

However, science is revealing that plants can sense and respond to a variety of environmental (both biotic and abiotic) stimuli as they move through surrounding environments. Sensory capabilities allow plants to develop specific growth responses to react to external changing conditions. It is noteworthy that plants can communicate information to other plants, animals or even fungi, as well as implement complex, adaptive behaviors.

To design the first plant-inspired robot, specific experiments were designed to investigate root penetration capabilities and the kinematics of root movements, as well as the tropism mechanisms and root behavior in controlled and stress conditions. The results of these scientific experiments provided the benchmark data to build the Plantoid robot, its sensory-actuation mechanisms and the architectural behavior. The Plantoid benefits from cutting-edge data-gathering and data-modeling techniques as applied to plants. Its functioning requires an interdisciplinary merging of various bodies of knowledge (including biology, material science, engineering and robotics), which goes far beyond the typical boundaries of a traditional robotic manufacturing approach.

Thus, the Plantoid can be used as educational platform for allowing new generations to better understand the functionalities of real plants. A plant-savvy robot can educate humans to seeing plants as complex organism endowed with sensing capabilities, a peculiar form of intelligence, and unparalled adaptive capabilities. This ought to increase our respect for them.

Acknowledgements

This study was partially founded by the PLANTOID project (EU-FP7-FETOpen grant n. 29343).

Notes

1. New direction in plant ethics. Available at https://plantethics.univie.ac.at/ (accessed September 2017).
2. Online etymology dictionary. Available at www.etymonline.com/ (accessed September 2017).
3. The first time the word robot ethics appeared was in 1995 about a robot agent, see Lieberman 1995.
4. The First International Symposium on ROBOETHICS: The ethics, social, humanitarian and ecological aspects of Robotics' was organized by Scuola di Robotica in collaboration with Scuola Superiore Sant'Anna, the International Institute of Humanitarian Law and the School of Theology, Pontifical University of the Holy Cross, Rome, Italy and took place on 30th – 31th January 2004, in Villa Nobel, Sanremo, Italy.
5. Gianmarco Veruggio – who is considered as one of its founders – describes roboethics as "an applied ethics whose objective is to develop scientific/cultural/technical tools that can be share by different social groups and beliefs. These tools aim to promote and encourage the development of robotics for the advancement of human society and individuals and to help preventing misuse against humankind (Veruggio and Operto 2009).
6. For example, the prime yearly robotics conference, the IEEE International Conference on Robotics and Automation, featured seven roboethics workshops in the period 2005–2014.
7. www.ieee-ras.org/robot-ethics.
8. According to René von Schomberg RRI is "a transparent, interactive process by which societal actors and innovators become mutually responsive to each other with a view on the (ethical) acceptability, sustainability and societal desirability of the innovation process and its marketable products (in order to allow a proper embedding of scientific and technological advances in our society)" (von Schomberg 2011).

References

Banks, M. R., Willoughby, L. M. and Banks, W. A. 2008. Animal-assisted therapy and loneliness in nursing homes: Use of robotic versus living dogs. *Journal of the American Medical Directors Association* 9, no. 3: 173–7.

Bhushan, B. 2009. Biomimetics: Lessons from nature—An overview. *Philosophical Transactions. Series A, Mathematical, Physical, and Engineering Sciences* 367, no. 1893: 1445–86.

Bicchi, A., M. A. Peshkin, and J. E. Colgate. 2008. Safety for physical human-robot interaction. Pp. 1335–48 in *Springer handbook of robotics*, edited by B. Siciliano and O. Khatib. Heidelberg: Springer.

Collingridge, D. 1980. *The social control of technology*. London, UK: St. Martin's Press.

Hart, J. W. 1990. *Plant tropisms: And other growth movements*. London, UK: Unwin Hyman.

Ijspeert, A. J. 2014. Biorobotics: Using robots to emulate and investigate agile locomotion. *Science* 346, no. 6206, 196–203.

King, A. 2017. Technology: The future of agriculture. *Nature* 544, no. 7651: S21–3.

Lieberman H. 1995. Letizia: An agent that assists web browsing. *AAAI Technical Report* FS-95–03, 1995, AAAI. www.aaai.org.

Mazur, M. 2016. Six ways drones are revolutionizing agriculture. *MIT Technology Review*. Accessed September 2017 from www.technologyreview.com/s/601935/six-ways-drones-are-revolutionizing-agriculture/.

Mazzolai, B. 2017. Plant-inspired growing robots. Pp. 57–63 in *Soft robotics: Trends, applications and challenges. Proceedings of the Soft Robotics Week, April 25–30, 2016, Livorno, Italy*, edited by C. Laschi, J. Rossiter, F. Iida, M. Cianchetti and L. Margheri. Cham: Springer.

Mazzolai, B., Beccai, L., and V. Mattoli. 2014. Plants as model in biomimetics and biorobotics: New perspectives. *Frontiers in Bioengineering and Biotechnology* 2: 2.

Noyes, J., Cook, M. and Masakowshi Y. (ed.). (2007). Decision making in complex environments. Aldershot UK: Ashgate Publishing Limited.

Parasuraman, R., Molloy, R. and Singh, I. 1993. Performance consequences of automation-induced complacency. *The International Journal of Aviation Psychology*, 3, no. 1, 1–23.

Pfeifer, R., Lungarella, M. and Iida, F. 2007. Self-organization, embodiment, and biologically inspired robotics. *Science* 318, no. 5853, 1088–93.

Sadeghi, A., Mondini, A., Del Dottore, E., Mattoli, V., Beccai, L., Taccola, S., Lucarotti, C., Totaro, M. and Mazzolai, B. 2016. A plant-inspired robot with soft differential bending capabilities. *Bioinspiration & Biomimetics* 12, no. 1: 015001.

Sadeghi, A., Mondini, A. and Mazzolai, B. 2017. Toward self-growing soft robots inspired by plant roots and based on additive manufacturing technologies. *Soft Robotics* 4, no. 3: 211–23.

Taccola, S., Greco, F., Sinibaldi, E., Mondini, A., Mazzolai, B. and Mattoli, V. 2015. Toward a new generation of electrically controllable hygromorphic soft actuators. *Advanced Materials* 27 1668–75.

Turkle, S. 2007. Authenticity in the age of digital companions. *Interaction Studies. Social Behaviour and Communication in Biological and Artificial Systems* 8, no. 3: 501–17.

Turkle, S. 2012. *Alone together: Why we expect more from technology and less from each other*. New York, NY: Basic Books.

Verbelen, J. P., De Cnodder, T., Le, J., Vissenberg, K. and Baluška, F. 2006. The root apex of arabidopsis thaliana consists of four distinct zones of growth activities. *Plant Signaling Behavior* 1: 296–304.

Veruggio, G., and Operto, F. 2008. Roboethics: Social and ethical implications of robotics. Pp. 1499–524 in *Springer handbook of robotics*, edited by B. Siciliano and O. Khatib. Berlin, Germany: Springer-Verlag.

von Schomberg, R. 2011. "Towards Responsible Research and Innovation in the Information and Communication Technologies and Security Technologies Fields." Pp. 7–16 in *Towards Responsible Research and Innovation in the Information and Communication Technologies and Security Technologies Fields*, edited by R. von Schomberg. Luxembourg: Publications Office of the European Union.

Wong, T. S., Kang, S. H., Tang, S. K. Y., Smythe, E. J., Hatton, B. J., Grinthal, A. and Aizenberg, J. 2011. Bioinspired self-repairing slippery surfaces with pressure-stable omniphobicity. *Nature* 477, no. 7365: 443–7.

Yang, G.-Z., Bellingham, J., Choset, H., Dario, P., Fischer, P., Fukuda, T., Jacobstein, N., Nelson, B., Veloso, M. and Berg, J. 2006. "Science for robotics and robotics for science." *Science Robotics* 1, no. 1., eaal2099, DOI: 10.1126/scirobotics.aal2099.

Zang, J., Ryu, S., Pugno, N., Wang, Q., Tu, Q., Buehler, M. J., and Zhao, X. 2013. Multifunctionality and control of the crumpling and unfolding of large-area graphene. *Nature Materials* 12, no. 4: 321–5.

Index

Note: Italic page numbers refer to figures and page numbers followed by "n" refer to endnotes.

For Product Safety Concerns and Information please contact our EU representative GPSR@taylorandfrancis.com
Taylor & Francis Verlag GmbH, Kaufingerstraße 24, 80331 München, Germany

www.ingramcontent.com/pod-product-compliance
Ingram Content Group UK Ltd.
Pitfield, Milton Keynes, MK11 3LW, UK
UKHW011002180926
475701UK00004B/525
9780367855994